THE NEW COSMIC STORY

THE NEW COSMIC STORY

Inside Our Awakening Universe

JOHN F. HAUGHT

Yale

UNIVERSITY PRESS

New Haven and London

Published with assistance from the Mary Cady Tew Memorial Fund.

Yale University Press books may be purchased in quantity for
educational, business, or promotional use. For information,
please e-mail sales.press@yale.edu (U.S. office) or sales@yaleup.co.uk
(U.K. office).

Set in Janson Text type by Integrated Publishing Solutions,
Grand Rapids, Michigan
Printed in the United States of America.

Library of Congress Control Number: 2017930761
ISBN 978-0-300-21703-2 (hardcover : alk. paper)

A catalogue record for this book is available from the British Library.

This paper meets the requirements of ANSI/NISO Z39.48-1992
(Permanence of Paper).

10 9 8 7 6 5 4 3 2 1

To my grandchildren
Christopher, Caitlyn, Dominic, Colin, and Jacob

Contents

Acknowledgments   ix

INTRODUCTION   1

1. Dawning   7
2. Awakening   27
3. Transformation   49
4. Interiority   65
5. Indestructibility   79
6. Transcendence   93
7. Symbolism   111
8. Purpose   127
9. Obligation   143
10. Wrongness   159
11. Happiness   175
12. Prayerfulness   189

Notes   203
Index   221

Acknowledgments

I WISH TO THANK John Loudon for his encouragement and support throughout the writing of this book. It has been a pleasure working with him and benefiting from his wisdom. Thanks also to editors Heather Gold and Dan Heaton. I wish that every writer could have as perceptive and learned an editor as Dan has been in this book's production. Thanks also to Philip Novak, Charles A. O'Connor III, Robert Ulanowicz, Elizabeth McKeown, and especially my wife, Evelyn, for reading the manuscript, for their generous encouragement, and for their helpful suggestions. Thanks, finally, to the American Teilhard Association for the opportunity to deliver the annual Teilhard Association Lecture in New York in 2014 and to share with its members some of the ideas adapted here in Chapter 4.

Introduction

OVER THE PAST TWO centuries scientists have found out that the universe is a story still being told. During the past hundred years they have learned that our Big Bang universe began billions of years before life appeared and even more billions before humans arrived on planet Earth. New scientific awareness of the long cosmic preamble to human history has inspired attempts recently to connect the relatively short span of our own existence to the larger cosmic epic. Sometimes these efforts are referred to as Big History. Big History seeks, as best it can, to tell the story of *everything* that has taken place in the past, including what was going on in the universe before Homo sapiens arrived.

How consequential Big History will turn out to be in the long run is debatable. So far most versions have stapled the human story only loosely onto scientific accounts of the earlier cosmological and biological chapters. They have seldom looked deeply into how one stage interpenetrates the others.[1] Books on Big History typically start off with several chapters summarizing material that readers could pick up from any good popularization of scientific discoveries. Then they follow this up with content that any reliable summary of human history has already made available. Missing from their extended narratives is a dramatic interlacing of the various epochs. Above all, what is lacking is sustained reference to what I will be calling the inside story of the universe.

Big History scholars locate—and deflate—the human story by

placing it against the backdrop of the universe's spatial and temporal immensity. This is a useful point of view, but not the only one. The universe, after all, includes *subjects*, hidden centers of experience whose significance cannot be measured by science or captured by purely historical reporting. What is needed, I believe, is a narrative that tells the whole cosmic story, inside as well as outside. Startlingly absent from Big History so far, for example, is a sense of how religion fits into the cosmic story. This book is an attempt to address this omission. In it I argue that we cannot expect to understand well what is going on in cosmic history apart from a careful examination of what goes on in the interior striving of life that reaches the summit of its intensity in humanity's spiritual adventures.

Religious experience is part of the inside story of the universe. For thousands of years religious sentiments have come down from one human generation to the next packaged in symbolic forms whose meaning is mostly inaccessible to science. Yet the emergence of religious subjectivity, though hidden, is just as much part of the universe as is the formation of atoms and galaxies. Big History, following the methods of physical science, characteristically pays little serious attention to religion and other things going on inside. My objective here then is to look inside the story of the universe without ignoring the outside. I am convinced that any big history that lives up to its name needs to correlate what is going on objectively with what is going on subjectively.

Subjectivity, as far as we can tell, burns most feverishly in humans. It has also been emerging more quietly in the story of life—and implicitly throughout the whole cosmic journey—for billions of years prior to our own recent arrival. The cool detachment of science, however, never feels fully the heat of inner experience and the dramatic quality of its emergence. I believe then that the scientific approach idealized by Big History needs to be supplemented by a wider attentiveness and a more sweeping empiricism than science usually employs. A really big history must take into account the interior dimension of living, thinking, and worshiping subjects and not just outward, publicly available events.

Big History is aware of religion externally as a social and cultural phenomenon and as a motor for historical change, but its outside perspective passes over religion's dramatic inner substance. Along with many other contemporary intellectuals, authors of Big History usually assume that science provides the widest and most reliable pathway to knowledge. A sizable number of them take for granted that the world available to science is really all there is. Subjectivity in that case is scarcely distinguishable from nothingness, and religion is no more than a filmy human concoction that evaporates altogether when subjected to scientific examination.

From the start, however, the cosmic story has carried with it, at least faintly, a scientifically inaccessible lining of "insideness." As we shall see, the cosmos *is* in fact a story of emerging interiority. In the case of humans, subjectivity has become palpably manifest in our many passions, our sense of freedom, ethical aspiration, and aesthetic sensitivity, but especially in our longing for meaning and truth. So hidden is this interior dimension that scientific materialists—for whom measurable "matter" is all there is—are unable or unwilling to tell the whole cosmic story. They pass over in silence the most interesting aspect of the drama—namely, the emergence of an interior world consisting of sentience, intelligence, moral aspiration, and religious passion. Materialist thinkers generally ignore the fact that nature includes an undeniably real inside story composed of struggling subjects who yearn for a kind of satisfaction that science's outside point of view fails even to notice, let alone understand.[2]

The epic of the universe, I argue, is no less a story about emerging subjectivity than about the movement of atoms, molecules, cells, and social groups. With the relatively recent arrival of distinctly religious experience in cosmic history, the universe reaches out toward horizons previously unknown. If Big History were to include in its survey the totality of events in cosmic time, it would have to find a way to connect the outer sequence of physical events to the drama going on inside. In a special way, it would have to ask what religion tells us about the universe out of which it has recently come to birth.

My objective, then, is to reflect on the cosmic meaning of religion as well as on the religious meaning of the cosmos. Before beginning to do so, however, I need to say a word up front about the function religion has played in *human* life from the beginning. In the broadest sense, along with the British scholar John Bowker, I take religion to be the primary way in which people have sought pathways through the severest limits on life.[3] These limits include not only the threat of death but also the experience of fate, guilt, doubt, and meaninglessness. Religion, since its earliest beginnings, has been a highly symbolic search for something permanently trustworthy, something we can always rely on to give us the courage to conquer the anxiety that comes with being finite, striving, and mortal beings.[4]

Religion has also been the main way in which people have sought final deliverance from suffering. In the Hindu Upanishads, to give only one example, we read that suffering may be conquered if we can overcome the illusion of existing separately from *Brahman*, that is, Infinite Being, Consciousness, and Bliss. If we could only realize that the temporal world in which suffering exists is illusory, then we would not have to let suffering triumph over us. Suffering can defeat us only as long as we think, in our typical state of ignorance (*avidya*), that the world of appearances (*samsara*) is the real world. The Upanishads instruct us, however, that the domain of suffering is not real. Suffering may seem unconquerable, but only so long as we are under the illusion (*maya*) that our existence is separate from that of Brahman. The function of religious meditation and practices (*yoga*) is therefore to bring us to a state of liberation (*moksha*) from samsara and hence from suffering.

Once we get past the illusion of our separateness from Brahman, we may not immediately cease to feel suffering, but, as Bowker writes,

> the individual who has an adequate grasp of Brahman will find that suffering falls away in significance. Since everything that happens is a manifestation of Brahman, it follows that true understanding only arises when the accidents of time and space are penetrated and are seen to reveal Brahman. Brahman pervades all things without being exhausted in any one of them; which means that suffering or sorrow cannot be the final truth about existence.[5]

The Upanishads illustrate the more general religious assumption that suffering can be overcome by an awakening and transformation that allow us to see things rightly. Religion, however, has to do not only with the need for consolation and healing in the face of perishing and suffering but also with the overflowing sense of wonder at the fact that anything exists at all. In this respect religion has its origin in a sense of grateful surprise at the mystery of being. At some level, all conscious beings, including those who call themselves irreligious, experience the shock that anything exists at all. We humans, however, have devised countless ways to avoid acknowledging the mystery of it all, today perhaps more than ever. In most eras of human history, nevertheless, responsiveness to the gift of existence has manifested itself in an instinct to worship a hidden and indestructible source of all being. This religious inclination has come to expression in symbols, analogies, metaphors, rituals, myths, and theologies. These obscure modes of communication point allegedly to an indestructible and transcendent dimension of being from which we came, toward which we are destined, and in whose ambience we find both moral guidance and a meaning for our lives. I want to ask what this phenomenon we call religion means in the context of our new scientific story of the universe.

In obedience to the assumption that only an "outside" approach can lead to right understanding, exponents of Big History are telling the cosmic story today without much looking inside, especially at religious experience. I intend to make up, at least modestly, for the general failure of scholars to include the emergence of religion in their surveys of cosmic history. In each chapter I therefore focus on one of twelve aspects of religion common to many traditions, asking what each distinct trait means in the context of an unfinished universe. I note also that every facet of religion treated here is now undergoing unprecedented efforts at debunking in the intellectual world, usually in the name of science.

The new picture of a universe that is still emerging allows us to understand in an unfamiliar way not only the lofty aspirations but also the undeniable evils in religion as highlighted recently, for example, by the New Atheists. By situating religion in the setting of an unfinished universe we may learn that, along with the suffering of life and the darkness of human experience in general, the wrongness in religion is a signal that the universe is still far from being fully actualized.[6] I do not mention on every page that religion at times gets mixed up with monstrous evil, but I know this to be the case. I assume that religion, since it is webbed into an unfinished universe, is unfinished too. Our new sense of the universe as an incomplete process of becoming allows us, now more than ever before, to confess frankly that religion, like everything else in our half-born world, has a shadow side.

There is always a risk, of course, in speaking of religion in the sweeping way that I do in this volume. The obvious drawback is that such a generalized approach fails to do justice to the particularity of each distinct tradition. The word "religion" itself is an abstraction that tends to gloss over not only the differences among communities of faith but also the unrepeatable way in which each person's interior life intersects with his or her tradition.[7] Nevertheless, abstractions are essential to any quest for intelligibility. This is certainly the case with the natural sciences, whose progress in modern times corresponds to the discovery of general laws that unify previously disconnected areas of research. Something similar is to be gained also in the search for an encompassing point of view that highlights characteristics common to many religious perspectives.

Various traditions, I believe, share traits that allow us to use the word "religion" meaningfully in referring to them all at once. As a Roman Catholic with ecumenical and interreligious interests, I am aware that some Christians, especially those who follow the Swiss Protestant theologian Karl Barth, distinguish sharply between "Christian faith" and religion. They accept the first and distrust the second. Exclusivist comparisons, however, make no sense apart from a tacit agreement on the general topography of the territory being contested. Religious traditions are not all saying the same thing, but even with all their differences they have common interests and dispositions worth highlighting. They all assume, for example, the existence of an interior life and of the need to undergo awakening and transformation. They nourish a sense of obligation, and they all idealize "rightness." They speak symbolically, mythically, and metaphorically about evil, perishing, purpose, everlastingness, happiness, and transcendence. Only against the backdrop of these constants do the variables among religious traditions show up at all. By situating the common attributes of religion inside a universe that is still emerging, we may come to see all of them in a whole new light.

CHAPTER ONE

Dawning

And though the last lights off the black West went,

Oh, morning, at the brown brink eastward springs—

—GERARD MANLEY HOPKINS

The darkness is passing away, and the true light is already shining.

—I JOHN 2:8

ONLY RECENTLY HAS SCIENCE demonstrated beyond reasonable doubt that the universe is still coming into being, that nature is narrative to the core, and that the cosmic story is far from finished. Thanks to developments especially in geology, biology, and cosmology we now know that we live in a universe that is still on a long journey. And even though we cannot see where it is going or exactly how it will end, nothing in the cosmos now looks the same as before, including religion.

Formerly nobody would have made such a claim. Religious believers usually assumed that the natural world is a fixed stage for the human drama. Philosophers did too. Immanuel Kant (1724–1804), for example, thought of the physical universe not as something to focus on, but as a backdrop for the human quest for meaning, truth,

7

and goodness. The universe to this great thinker was mostly a setting for the human pursuit of enlightenment, moral integrity, and God. The three most important questions we can ask, he said, are: What can I know? What must I do? And what may I hope for? If he were with us today, however, Kant would surely add a fourth: What is going on? Informed by today's science, he would realize that the universe is not a stage but the whole show. He would wonder not only about the stars above and the moral law within, but also about whether anything of lasting significance is working itself out in the universe.

During the past two centuries, as a gift of science, we have felt the natural world moving beneath our feet. We now realize that it has never stayed the same for long. After tossing and turning for billions of years, the universe is now waking up. As industrialists dug for coal during the nineteenth century, they disgorged from the hillsides a fossil record of life becoming increasingly alert over long spans of time. Since its first appearance life on our planet has taken about 3.8 billion years to get where it is today, suffering through five major extinctions millions of years apart. Unplanned events in natural history have sent life scrambling off in different directions and on many paths, including the winding way that led to us.

Yet even before life and human consciousness came along, as cosmologists have recently confirmed, the whole universe was already a dramatic performance. Astrophysicists have tracked life's prelude all the way back to the beginning of the Big Bang universe. And the journey continues. If we think seriously about our unfinished cosmos, as I intend to do in this book, we shall have to entertain new thoughts about everything, about who we are and where we are going, and about the meaning of our lives. Religion, whose business it has been to deal with these big questions, must also mean something new.

We humans are latecomers in the cosmic story, and religion came along with us. Humanlike ancestors showed up several million years ago, and we can tell from their tools that they could think and act purposively. They had an interior life and sought significance. We do not know for certain, but religion—the most important way in which people have looked for meaning—probably arose at least indistinctly with the birth of human consciousness. For better or worse, humans arrived already prepared to look beyond the limits of their lives. Modern humans, as nearly as we can tell, were already migrating out of Africa 200,000 years ago. By 30,000 years ago they were painting on cave walls, carving figurines, and marking graves. By that time humans had acquired interior lives and linguistic skills rooted in a talent for symbolic representation. Without their extraordinary ability to signify one thing symbolically by pointing to something else, neither language nor religion would ever have come about in the cosmic story. Apart from symbolic consciousness there would be no thoughts of gods and no fervent expectations of final victory of life over death.

We may infer from its symbolic forms that religion is a set of sentiments and aspirations arising from deep inside human consciousness. Ever since the beginning of life on our planet the temperature of inner experience—what I am calling subjectivity—has continued to climb. As nerve tissue became increasingly complex in life's evolution, sentience emerged, emotion deepened, and eventually symbolism and thought broke out. Shortly after the birth of self-conscious minds, our ancestors began to develop the sense of a spiritual world and sought to encounter it by engaging in religious rituals. Religion, as John Bowker succinctly puts it, has always been a kind of route-finding that seeks pathways beyond the most intractable limitations on life.[1] Humans may be resourceful enough to carve out a limited living space through their own efforts, but their willful acts cannot vanquish the mystery that encompasses their lives. Nor can they simply wish away the threats of death, doubt, guilt, and meaninglessness. Since these barriers have yet to be removed, it is not surprising that most people on our planet, including the majority of educated people, are still religious.

Religiousness, as much as any other human quality, still sets our species sharply apart from all others. Nonhuman animals play, but they do not pray. When they are well fed, they go to sleep. When humans are well fed, they ask questions, one of which is whether their lives have a meaning.[2] We now wonder, however, whether something meaningful is going on not just in our personal lives, but also in the cosmic story to which science is connecting us more intimately with each new discovery.

This book represents an attempt to understand religion in the context of our new scientific story of the universe. The arrival of religion is one of the most intriguing developments not only in human but also in cosmic history. Yet the significance of religion for our understanding of the universe remains largely unexplored. Students of Big History are usually content to treat religion as part of the human story and hence as a topic to be explored by historical methods, the social sciences, and psychology. Religion, they also agree, is part of the story of life, and so it needs to be examined from the point of view of evolutionary biology as well. Here, though, I want to ask what religion means if we widen our perspective and examine it as part of an ongoing *cosmic* story.

Between two thousand and three thousand years ago a shift in human consciousness began to occur on Earth that was so unprecedented that it amounts to nothing less than a major new chapter in the history of the universe. Over a period of several centuries, especially in China, India, Europe, and the Near East, the religious quest for meaning became less symbolic and more mystical and theoretical than earlier. In the teachings of a few exceptional seekers and their followers, religion in these places became less concerned with rituals, petitions, and appeasement of supernatural beings and more preoccupied with personal awakening and spiritual transformation. Without rejecting popular piety altogether, new religious visionaries at the time began to cultivate the impression that an indestructible dimension of being, goodness, truth, beauty, and unity lies hidden beyond, or deep within, the world of ordinary experience. The purpose of our lives, they taught, is to awaken to this hidden realm of being and allow our lives to be transformed by it.

Karl Jaspers, a respected German philosopher (1883–1969), called this time of religious ferment the "axial age" (800–300 BCE). It was a period, he observed, during which "the spiritual foundations of humanity were laid simultaneously and independently in China, India, Persia, Judea, and Greece." These, he went on to say, "are the foundations upon which humanity still exists today."[3] It is roughly this same epoch of awakening that I have in mind when I refer to religion in this book. I intend to specify more carefully what I mean by the term as we move along, but I want to suggest even now that every aspect of religion gains new meaning and importance once we link it to the new scientific story of an unfinished universe. Each chapter in this book therefore takes a feature common to the axial religious traditions and explores its cosmic significance, a connection to which Jaspers and other scholars in the first half of the twentieth century could scarcely have given much thought.

The earlier lack of scholarly attention to the cosmic significance of religion is understandable since only in the latter half of the twentieth century did it become clear to most scientists and other intellectuals that the universe as a whole has a highly nuanced narrative character. Going beyond Jaspers and most other scholars of religion, therefore, I want to focus on what religion means in the context of a cosmic drama. In the Dao de Jing in China, in the Upanishads and the sermons of the Buddha in India, in the philosophical dialogues of Plato in Greece, and in the Abrahamic traditions of the Near East, a new turn was taking place in the story of the universe. As we shall see, it is not enough to view the extraordinary axial transformation of religious consciousness simply as a set of local geographical, historical, cultural, and terrestrial curiosities. It is also a whole new era in cosmic history.[4]

What was occurring during the axial period—and continues now—was the birth of a new sense of rightness. The new wave of consciousness began to make sharper distinctions than ever before between a right way and a wrong way to live, think, act, work, and pray. Indian mystics during the axial period, for example, distinguished a higher calling to reality and truth from a lower and lazier contentment with illusion and attachment to immediacy. They sought to purify piety of contamination by distracting symbolic imagery and warned against a life of vain attachment to passing allurements. Ultimate rightness, they said, is *neti neti*, "not this, not that." We cannot comprehend absolute rightness, but it can comprehend us. The Buddha (circa 500 BCE), in his Noble Eightfold Path, sought to teach *right* wisdom, *right* action, and *right* appreciation. Even though he was not concerned with finding a deity to worship, or a permanence beneath perishing, the Buddha was nonetheless measuring human piety, moral conduct, and wisdom in accordance with an incorruptible standard of "rightness." In China Laozi was looking for the right Way. In the Greek world Socrates and Plato noted the sharp difference between opinion and truth, between what is transient and imperfect on the one hand and what is real and perfectly good on the other. The prophets of Israel a bit earlier had laid out a path for authentic human existence in which "doing right" (*tzedek, tzedekah, mishpat*) came to be associated less with sacrificial offerings and more intimately with the ideal of social justice as commanded by a God whose preeminent concern was for the poor and oppressed. The prophet Micah taught that the right way to live is "to do justice, and to love kindness, and to walk humbly with your God" (Micah 6:8). Centuries later Jesus of Nazareth, identifying himself with the prophetic tradition, distinguished between the present age of injustice and the "right" age of compassion and peace now dawning. He spoke of a "reign of God" in which superabundant love would transcend mere fairness. His apostle Paul referred to the work of Jesus as "justification," a term that implies "making things right." The evangelist Luke wrote of the early Christian movement as "the (right) way" (*hodos*), and the Gospel of John presents Jesus as "the *way*, the truth and the life." Still later Muhammad would set forth the Five Pillars of Islam as the way to keep his followers on the right course in their pilgrimage on Earth.

When I use the term "religion" in this book, I therefore mean an awakening to the dawning of "rightness"—I can think of no more inclusive term—seeds of which were sown early in human history but that came to fuller flowering in the general time period, geographical places, and historical developments I have just indicated. It was during this "axial" time in human history that devotees began to acknowledge formally that rightness is real—indeed more real than anything else. What I want to emphasize is that the dawning of rightness was not just a set of interior human intuitions but also a great event in the history of the universe.[5]

An implicit attraction to rightness is intrinsic to human consciousness, but Jaspers turns our attention to a period of history when rightness came to be seen more explicitly as transcendent, indestructible, fulfilling, and liberating. He is right to point out that this period laid the foundation of contemporary thought and culture. Even the rationalism, scientism, and humanism in modern secular cultures appeal tacitly to the ideal of rightness—right thinking, right method, right doing—to which humans long ago awakened most explicitly by way of religion. When contemporary New Atheists focus on all the wrongness in religion, as they are certainly entitled to do, they are blissfully unaware that they too are indebted to an idealizing of rightness that first made its way into human cultures by way of religious consciousness many centuries ago.

When I speak of "religion" (a term I use deliberately in the singular), I do not want to leave out the many thousands of years of symbolic, devotional, and ritualistic practices that preceded the great awakening to which I am referring and that still persist in popular religion. Indeed, the sense of rightness on which this book focuses began to emerge at least faintly in preaxial religion, and it exists to some degree in spiritual visions everywhere. Nor do I want to suppress awareness of the holy horrors that have accompanied religion throughout its ragged history. Nevertheless, my emphasis is on religion as the dramatic arrival in the universe of a new kind of consciousness centered on rightness.

Axial religion is special for its idealizing a unifying principle of meaning, goodness, beauty, and truth, sometimes called God. This principle of unity, more often than not, is thought of as lying beyond the diversity and disintegration occurring in nature and human experience. As such, religion is partly responsible for generating and supporting the very idea of a *uni*-verse. Moreover, through the highly developed religious sense of a unifying principle of rightness, the universe has now reached a new awareness of its own internal unity. The very notion of a universe, in other words, owes much to the axial notion of a transcendent, consolidating principle of being and intelligibility. Whether religion has lived up to its own idealizing of unity, of course, is an important question, but here my attention is on the fact that through religion the whole cosmos has awakened from the inside to a mysterious horizon of integrating and indestructible rightness.[6] A refusal to acknowledge the full reality of this religious development amounts then to a serious impoverishment of our understanding of the universe.

Understanding religion as a new epoch in the unfolding of the cosmos is not the norm in religious studies. Yet the full drama of religious awakening would pass us by if we failed to look at it in terms of the impressive journey of the universe. If the physical world were merely a platform for the evolution of life, or a blank habitat for human history, religion could be dismissed summarily as a way for humans to escape from, or at best adapt to, an impersonal world.

Today, as we shall see in detail, this restrictive way of thinking about religion still permeates academic thought. Given that for countless modern and contemporary scholars the universe is pointless, it is not surprising that they have taken religion to be interesting fiction and empty gesturing at best. They are even embarrassed that so many of their fellow mortals have surrendered to religious naïveté instead of bowing to "reality" as science understands it. Intellectually elite observers still often regard religion as an annoying adhesion that needs to be surgically removed from both culture and cosmos—so as to leave more room for science. Contemporary evolutionists leniently grant that religion helped our ancestors survive long enough to bear offspring, so indirectly they concede that religion deserves our thanks. Nevertheless, they now consider it a distraction whose time is up.

In all this debunking of religion, a persistent background assumption is that the physical universe is pointless because nothing of lasting importance seems to be going on there. As it turns out, however, something *has* been going on in the universe, something undeniably and dramatically important. The drama began long before humans came along, and it is still in play. During the past two centuries, science has gradually presented the universe for all to see as a grand adventure, full of twists and turns we never knew about until recently. And today science allows ample room for even more surprises up ahead. As the cosmos has developed over billions of years, entirely new kinds of being—most notably life and thought—have emerged. These are two cosmic developments that none of us could have predicted had we been around to witness the inauspicious elemental state of the early universe. Since even more surprising developments may be waiting to take place farther along on the cosmic journey, therefore, the contemporary secularist verdict that the universe is manifestly pointless, that evolution is a meaningless experiment, and that religion is illusory, may be premature. The story of the universe, after all, is not over. For all we know, more impressive developments, some of them enabled by human technology, lie ahead. The universe, no matter how you look at it these days, is more than a stage for the evolution of life and a setting for human history. It is a continuing drama that keeps unlocking previously unpredicted possibilities. In doing so it opens up anew the ancient question: what is it all about?

A proper function of religion today, if it has any at all, is to address this question. What the universe is really all about, at least when read religiously, is awakening to the dawning of rightness. This dawning, however, is not without long shadows. Darkness remains even in religion. A cosmic approach allows us to understand better not only the religious sense of rightness but also the undeniable wrongness in religion. It takes into account the evils that confront every human life as well as those wrought by religious believers. A cosmic approach is fully aware of the exclusivity, inquisitions, crusades, misogyny, patriarchy, homophobia, dogmatism, violence, and killings that have been sponsored by religion on Earth. It takes for granted, however, that religion is inseparable from a universe still emerging, and hence that humanity's spiritual journey is not over. Because religion is a relatively new chapter in an ongoing cosmic epic, we should not be surprised that it is as unfinished as the universe that carries it. To demand its excision is therefore to terminate the cosmic story before it has a chance to play itself out. Contemporary attempts by prominent intellectuals to banish religion from human life and culture, as we shall see, are not only unreasonable but also cosmically catastrophic.

In telling the whole story of the universe, therefore, a truly big history can no longer plausibly leave out the world's recent awakening not only to life and thought but also, through religion, to an incorruptible and unifying principle of rightness. Religion is perhaps the most startling development so far in the inside story of the universe, but preparations for it were going on in natural history long before its actual arrival. Running silently through the heart of matter, a series of events that would flower into "subjectivity" has been part of the universe from the start. So hidden is this interior side of the cosmos from public examination that scientists and philosophers with materialist leanings usually claim it has no real existence. Leaving subjectivity out of their representations of the cosmos, however, they fail to tell the whole story. Consequently, if readers of the new cosmic story, such as those engaged in Big History, are serious about bringing deep coherence to their accounts, I believe they need to develop a more penetrating brand of inquiry, one that senses the inner drama and connects it tightly to the outer.

Adopting the outside perspective of scientific method, modern scientists and historians have become increasingly estranged from religion as they have lost touch with their own interiority. Scholars have become so accustomed to looking outward at the "objective" world that no inside story gets officially recorded in their portraits of reality. Contemporary high-minded repudiations of religion run parallel to—and are sustained by—the uncritical modern denial that subjectivity is part of the real world. Yet not only is subjectivity an undeniable aspect of nature, only through subjectivity can the universe finally acknowledge its own existence. Without subjectivity there would be no centers of experience to register the fact that anything exists or that it exists in a definite way. There would be no scientific inquiry, either. Moreover, if there were no real subjects, nothing would matter. It would be as though nothing even exists. Why there are subjects and not mere objects, and why the universe exists at all, may be two sides of a single question.

Subjectivity arose long before the arrival of humans, but its experiential and expressive power has intensified throughout life's evolution. In human beings it has reached a pitch unprecedented in the history of the universe, at least as far as we know. Human subjectivity includes not only bodily sensations and a wide range of psychic states but also the capacity to ask questions, to distinguish between right and wrong, to aim for beauty—and to yearn religiously for an imperishable rightness. Science has now taught us that the eventual emergence of subjectivity was already implicit in the physical properties of the early universe. The question of how life, mind, morality, aesthetic sensitivity, and religion came into being is now tightly connected to that of how the physical universe began and developed.

The origin of subjectivity somehow coincides with the birth of nature, as present-day astrophysics allows us to realize. Until roughly the past half-century, however, scholars knew little about the remarkable physical and astrophysical connections that exist between cosmos and consciousness (a linkage that I explore in more detail later). Scientific skeptics took for granted that the universe is fundamentally lifeless stuff suspended in space, perhaps forever. They assumed that, in the absence of a creator, mindless matter must be eternal and that sooner or later subjectivity would burst out only as a transient interruption of the universe's normal state of undisturbed silence. Remarkably, many scientists and philosophers still cling to this questionable set of beliefs. I intend, however, to expose the irrationality of the standard academically endorsed conviction that the universe is essentially mindless. I provide below both scientific and logical reasons to doubt the assumption that subjectivity is an eventually extinguishable anomaly in an underlying cosmic mindlessness.

If I am right about this, then the emergence of religious subjectivity is not an infantile escapism that humanity now needs to outgrow, nor is it an evolutionary adaptation to be condoned condescendingly. Rather, it is an indispensable new epoch in a cosmic story. I claim not that a cosmic perspective on religion is the only way to look at it but that our new scientific and historical awareness of an ongoing cosmic journey now offers us an unprecedented access to understanding the religious quest in a fresh way. I believe also that religious traditions the world over will become increasingly irrelevant, especially to educated people, unless they too start taking into account the new scientific story of the universe that gave birth to them. Again, there is much more to religion than cosmology allows us to see, but bringing up every aspect of it in a single book would be to lose focus. My purpose is not to provide a historical survey or comparative study of religious experience. Rather, I want to ask what religion, broadly speaking, means for our understanding of the universe, and what the new story of the universe means for our understanding of religion.

The Unity of Religious Experience

As long as the cosmos was taken to be essentially unchanging—an assumption that even the great Albert Einstein had cherished until his own equations proved him wrong—religion could easily seem to be the escapist illusions of a childish species refusing to reconcile itself to a seemingly stationary and impersonal habitat. Science, however, has now demonstrated that the universe is still on the move, and current cosmology allows us to interpret religious expectation as a relatively new stage in the universe's ongoing adventure. If the cosmos were fixed and finalized, as most humans had assumed until recently, it would be tempting to view religion as a flight *from* the world. But if the universe is still coming into being, religion can be thought of as a new phase in a cosmic passage. Religion is one of the main ways in which the cosmos, at least in its terrestrial precincts, not only seeks refuge from nothingness but also undergoes further intensification of its own being. From a cosmic-narrative point of view, therefore, the reckless contemporary intellectual discrediting of religion amounts to nothing less than a condemnation of the natural world to premature death.

Placing religion in the setting of our new cosmic story also allows for a fresh appreciation of the unity of religious experience and thus gives new significance to conversations among religious traditions in their common battle against cosmic pessimism. The real unity of religious traditions, I argue, does not lie in some shared primordial revelation that may have occurred long ago. Nor does it consist of a common evolutionary origin, a standard suite of psychological instincts and social habits, or kindred patterns of moral aspiration. Rather, the unity of religious traditions lies in their joint anticipation of a rightness that is now mostly out of range. In this respect religion's anticipatory leaning makes it remarkably confluent with the new scientific story of a universe that is still coming into being. Furthermore, different religious traditions, by becoming aware of their common setting in an unfinished universe, can begin to realize as never before that they may journey toward the elusive goal of rightness arm in arm with others.

Until now there have been four main ways of approaching the question of the unity of religion. The first is exclusivism. This is the belief by devotees of any particular religious tradition that they alone possess the one full and final revelation of truth. Embracing a single set of beliefs as unique and exclusive of others, accordingly, seems for them to be the only way to conquer the contradictions among the many different traditions. All other religious visions must be wrong or deficient when compared to the exclusivist's own sense of meaning and truth. Hence adherents to other traditions must undergo a process of conversion to the one legitimate faith if they expect to find the right route to final liberation and fulfillment.

A second way of seeking the unity of religious experience is known as perennialism. Expressed primarily in what is sometimes called the Perennial Philosophy, this venerable approach assumes that a single primordial revelation of truth was given to humankind either in a special revelatory moment beyond time, or in the distant historical past. According to perennialists, the initial rightness of this authoritative revelation became corrupted in the course of human history. The goal of religion at present, then, is to transport our hearts, minds, and souls back to the purity of the original revelation. Unlike exclusivism, however, perennialism highly values the present diversity and particularity of the world's religious traditions, viewing each of them as a legitimate way back to the pure illumination now sullied by the history of human ignorance and evil. Perennialists believe that each of the main religious traditions has the mission of pointing its followers toward the primordial revelation, so there is no need to pass over from one form of religion to another in order to get closer to eternal meaning and truth. Each tradition provides a pathway to the unalloyed original disclosure of rightness. Religious traditions are like many rivers flowing down a mountain from a common source at the summit. At the bottom of the mountain they stream away from one another, but they share a common fountainhead on high. Each tradition, consequently, is an imperfect but fruitful point of departure for making our way back upstream to the aboriginal font from which the many culturally distinct forms of devotion have sprung and then later split off.[7]

A third way of understanding the unity of religious experience is that of scientific naturalism. This is the belief that modern scientific method is the only right way to understand anything and that the physical universe is really all there is. Scientific naturalism treats all of religion as fiction traceable to a purely material, biological, or neurological set of natural occurrences. Like perennialism, it looks for unity in a common origin of all the diverse forms of religious experience. Since God does not exist and revelation is fiction, however, the common origin has to be purely natural. That is, there has to be at bottom a physical, scientifically accessible way of understanding what religion is and why it exists. Naturalistic psychologists and sociologists locate the underlying unity of religion in human drives and instincts. Today, though, the most confident naturalistic interpreters of religion try to explain it in evolutionary terms, as one of the ways in which our human ancestors adapted to a dangerous world. I shall have much more to say about evolutionary accounts of religion.[8]

A fourth approach to the question of religious unity is postmodern pluralism. This new outlook scorns all attempts to discover a single unifying theme in religious experience. It acknowledges the incurable diversity of traditions, so it casts aside all efforts to harmonize the various traditions either scientifically or theologically. As one recent pluralist puts it, "If practitioners of the world's religions are all mountain climbers, then they are on very different mountains, climbing very different peaks, and using very different tools and techniques in their ascents."[9] Pluralism rejects attempts to establish the superiority of one religious path over the others, and it despairs of ever reaching anything like the primordial truth that perennialism seeks. It also renounces attempts to make religion intelligible in terms of psychology, sociology, biology, and other academic disciplines. Pluralists are content to wade around in the irreducible multiplicity of religious experiences and leave it at that. Unlike the approach I am taking, pluralism has no interest in connecting religion to the ongoing story of the universe.[10]

I propose, then, a fifth way of understanding the unity of religion. I call it the anticipatory approach. It assumes that religion is a relatively new chapter in the story of an emerging universe and that the emergence of religion, especially as epitomized in the axial traditions, is a no less momentous development in the cosmic story than the earlier arrival of life and mind. When life came along, and then more recently when mind emerged, the universe became dramatically different from what it had been before. From the point of view of the elementary natural sciences, the appearance of life and mind brought nothing new into existence, since no physical laws were violated. From a cosmic perspective, however, life and mind were astonishing disturbances of the status quo. So too is the rise of religion. In and through religion the universe has finally begun to awaken consciously and gratefully both to its own emerging unity and to the dawning of a transcendent and indestructible rightness.

The unity of religious experience, from this fifth perspective, is a goal that has yet to be realized since the universe, of which religion is a part, is still in the process of becoming. The unity of religion, like that of the universe, can only be anticipated, not recalled. The quest for religious unity does not consist of pursuing a primordial revelation. Instead, the many religious traditions can discover a new kind of solidarity in a shared, though still diffuse, anticipation of a rightness rising on the horizon of an unfinished universe.

This dawning, moreover, may be thought of not as compelling but as inviting the universe toward new being from a horizon ahead. Rightness, from an anticipatory perspective, is just barely breaking over "the brown brink eastward," as the poet G. M. Hopkins names it. Rightness—a widely shared religious ideal and goal—is not simply a deposit handed on from an idealized past or given from on high. It can properly be approached only in an open and attentive spirit of expectation. Once they agree that rightness is reached by anticipation more than by recollection, religious traditions would have a new reason for mutual tolerance. Interaction among the traditions would then consist not of squabbling over access to forgotten truth but of asking how each may contribute to the ongoing awakening of the universe.

Of course, to agree on an anticipatory approach to religious unity, devotees of the various traditions would first have to open their minds to the recent scientific discovery of life's evolution and a 13.8 billion–year-old universe still in process of becoming. They are far from having done so. In some traditions and many sects there is heated opposition to science, and especially to evolutionary biology, just as there is unnecessary opposition to religion by scientific naturalists all over the world. However, if the arrival of religion is a new epoch in the cosmic journey, as I am proposing, in principle there is no reason to dispute its consonance with the new narrative scientific portrait of the universe. Matter, life, mind, and religion are successive emergent epochs in a single cosmic story. And, as I shall argue, the unlimited rightness to which religion makes reference is not opposed to science but is instead a goal that may quietly motivate the scientific search for right understanding.

The anticipatory approach I am laying out here also allows religious believers to accept the relativity of their native or chosen pathways without requiring any capitulation to relativism. Historical and cultural relativity does indeed limit each tradition, but it need not diminish each tradition's confidence in the absoluteness of the goal toward which it points. A particular route may anticipate what is absolute, but it does so from within a particular cultural, symbolic, linguistic, and historical situation. As long as allowance is made for both the unfinished character of the cosmos and the anticipatory nature of religion, the unity on which both converge cannot possibly be an exclusive current possession of any one tradition. Yet it may be the aim of all.

Furthermore, by taking into account the results of modern scientific inquiry, an anticipatory approach finds no room for real conflict between science and religion. Science, of course, has a relativity of its own, as every good scientist is willing to admit. Yet some scientific discoveries are so far beyond dispute that it would be silly to ignore them as mere guesses. One of these is the realization that the universe is a work in progress, a fact now as unassailable as our awareness that the earth revolves around the sun. Denial of the discoveries of biology and cosmology is both intellectually and religiously fatal. Whenever religious teachers ignore scientific discoveries, they are making way for the eventual extinction of their own traditions. To have a meaningful future, religion all over the world needs now to come to grips with the new scientific understanding of the natural world.

The "exclusivist" solution to the problem of religious plurality, in that case, is especially unsustainable. Assuming an already completed universe, exclusivists typically take religion to be a performance humans carry out on a supposed cosmic stage where only one version of the play is right and all others wrong. Yet the universe's unfinished status affects everything it carries, including religion. In a still-emerging universe, each tradition needs to face the fact that it too is unfinished. Every religious idea or convention is to some degree exploratory rather than fixed or finalized in its letting in the light of rightness.

The perennialist notion of a perfect primordial revelation, consequently, is especially incompatible with our new scientific understanding of the universe. Any assumption that revelatory truth was delivered to humanity cleanly and tightly packaged long ago is irreconcilable with contemporary evolutionary biology, paleontology, history, archaeology, and cosmology. If the universe, as science has demonstrated unmistakably, is still on the way, religion is coupled irreversibly to the overall cosmic incompleteness. Religion on Earth, then, can reach integrative unity only by anticipation, not by recollection or Platonic participation "here below" in timeless truths allegedly enshrined eternally above. In an unfinished universe, religion is an adventurous anticipation of what is just beginning to dawn rather than a return to what has already been. Nor is it a basking in the glow of an eternal present. Faced with the diversity of religious traditions, the quest for unity is laudable, but this goal cannot lie in the discovery of a pristine past or an eternal completeness free from the cosmic conditions in which everything, including religion, is still coming to birth.

An unfinished universe offers a scientifically congenial point of departure for seeking the unity of religious experience. The current empirically certified picture of a still-emerging universe can direct both intellectual and religious attention toward what is yet to come rather than exclusively toward what is or what may have been. Religion in our new cosmic perspective still looks for a redemptive rightness and revelatory fullness, but it does so by expectancy rather than mere recovery. Instead of rejecting evolutionary science, as many religious believers feel compelled to do, we may now understand all religious quests as ways in which the cosmos, having recently become conscious of itself in human beings, anticipates deeper communion with the elusive rightness still dawning. Each religious tradition consists then of a unique way of approaching this goal. As I mentioned earlier, Buddhism's Noble Eightfold Path, for example, calls for right wisdom, right action, and right mindfulness. And even though Buddhism, unlike the Abrahamic traditions, does not associate rightness with divine personality, its idealizing of rightness aligns it nonetheless with the general religious awakening of the cosmos.

The anticipatory approach I am recommending also questions scientific naturalism's search for the unity of religion. Scientific naturalism seeks a unified account of religion by tracking it back, like every other set of natural phenomena, to a series of physically specifiable causes arising out of an aimless cosmic and evolutionary past. Naturalistic theories, however, fail to grasp the anticipatory leaning of religion because they fail, first of all, to grasp the anticipatory character of the cosmos itself. They understand religion, like all other living and mental phenomena, to be just one more instance of the world's original physical simplicity now "masquerading as complexity."[11] Since scientific materialism equates intelligibility with the reduction of everything to elemental physical simplicity, it seeks to understand religion also by trying to reveal its past physical causes. In subsequent chapters I argue that exclusively physicalist attempts to explain religion away end up unfortunately explaining science away also.

Finally, a word about the postmodern pluralist option. This approach is correctly respectful of religious differences, but it is a dead end both intellectually and spiritually. The human drive to understand cannot help but seek unity in the manifold data of experience, including religious experience. Nothing can be rendered intelligible apart from the search for a coherence that unifies particulars without overlooking their differences. This is true not only in the sciences—for example, in gravitational and evolutionary theory—but also in other areas of understanding. The quest for coherence, however, always entails the mind's moving toward ever more comprehensive points of view. This must be the case also in our trying to understand religion. By discouraging attempts to discover unity within the diversity of religious experience, strict pluralism violates a fundamental principle of human cognition, namely, the need for integration. Unfortunately, it also fails to take advantage of the possibilities of understanding religion now made available by the discovery of an unfinished universe. I believe the anticipatory cosmic approach I am taking in this book offers a more open, unifying, and realistic point of departure for the study of religion than any of the other approaches just listed.

Religion and the Cosmic Future

Is religion, however, a good thing for the universe? Again, this is a question we could not have asked meaningfully until recently. Earlier in the modern age it seemed right to ask whether religion is good for human psyches, societies, and reason. And today biologically influenced inquirers ask whether religion serves any useful evolutionary function. But recent developments in astronomy, astrophysics, and ecology now allow us to ask—as never before—whether religion has cosmic importance. Until recently, modern scholars have approached religion through less expansive scientific and historical horizons of inquiry than that of cosmology. Often they have been content to interpret religion as a product of wishful thinking or as a flight from reason and science. Today, however, science allows us to view the whole cosmos as an unfinished story. In that setting, religion is not simply a human drama of salvation playing itself out on the platform of an indifferent physical world, but a new and still only half-born epoch in the career of the cosmos. Instead of ignoring and sometimes even sneering at religion, serious thinkers, including students of Big History, may henceforth understand the human aspiring to rightness as crucial to the coming-to-be of the universe. Wide-minded scholarship in touch with the new story of the natural world will seek to nurture and purify religion rather than wipe it away. Furthermore, if the universe is itself a birth, we have every reason to anticipate further stages of awakening in its future coming to be. Religion, I suspect, is likely always to be an indispensable component of this great drama.

CHAPTER TWO

Awakening

In nature every moment is new; the past is always swallowed and forgotten; the coming only is sacred.

—RALPH WALDO EMERSON

THE LONG AND GRADUAL cosmic awakening to rightness has become explicit, in widely different ways, in wisdom traditions such as Hinduism, Daoism, Buddhism, Jainism, Judaism, Zoroastrianism, Platonism, Christianity, and Islam. By promoting values that they take to be universally and unconditionally good—for example, compassion for life, love of truth, and care for other humans—religious traditions sometimes think of rightness worshipfully. For them it is the transcendent ground of all being, truth, and value. At times they enshrine, adore, and even personify rightness, treasuring it as what is most real. To emphasize its universal and inexpressible realness, they endow it with the qualities of hiddenness, transcendence, and indestructibility. To religion the inaccessible, comprehensive, and unsettling reality of rightness is the ultimate reason why humans seek truth, why we have a sense of obligation, and why we long restlessly for perfect beauty. In our intellectual, moral, and aesthetic experience we all, at least tacitly, anticipate rightness—even when we deny it. This becomes obvious whenever we catch ourselves in the act of looking for right understanding, right action, and right satisfaction. Even ordinary human consciousness is inseparable from a tasting of rightness.

In religion, however, the tasting intensifies to the point of savoring. In religion people become explicitly and thankfully aware of the reality of rightness. Religion is a gradual but grateful awakening to the elusive horizon of unrestricted being, goodness, truth, and beauty. These are "transcendental" ideals that, for the sake of linguistic economy, I refer to collectively as rightness. "Rightness" is a term that can stand for what religion, generally speaking, esteems and anticipates. Religion, as the philosopher Alfred North Whitehead (1861–1947) writes, is

> the vision of something which stands beyond, behind, and within, the passing flux of immediate things; something which is real, and yet waiting to be realized; something which is a remote possibility, and yet the greatest of present facts; something that gives meaning to all that passes, and yet eludes apprehension; something whose possession is the final good, and yet is beyond all reach; something which is the ultimate ideal, and the hopeless quest.[1]

The "something" of which Whitehead speaks I refer to as rightness. From the point of view of religion, rightness is not an invention of the human imagination but something wider, deeper, more durable, more mysterious—and more hidden—than anything we experience in ordinary states of consciousness. Yet as our new sense of an unfinished universe allows us to appreciate today, there is a sense in which rightness is also something "waiting to be realized." Without a profound sense that there exists, already and forever, an indestructible rightness toward which the universe is turning, our sense of meaning, truth, goodness, and beauty would be groundless. Without a sense, however, that our own lives along with the whole cosmic process can contribute something new and indispensable to the "realizing" of rightness, these same lives—and the whole cosmic story—would lack significance.[2] Apart from the reality of indestructible rightness we would have no good reason to protest the wrongness in our lives, in the natural world, in human societies, and in religion too. Nor would we have any reason to anticipate final liberation from the negativities of existence. At the same time, however, if rightness were in every sense already fully realized, we might wonder what the point of an unfinished universe, along with the story of life and human existence, might be.

Because rightness is not already realized, moreover, there is inevitably an opening in the fabric of an unfinished universe not only for increasing growth but also for darkness and shadows, including in religion. Whitehead, who gives us the highly idealized definition of religion cited above, goes on to confess, quite rightly, that religion "has emerged into human experience mixed with the crudest fancies of barbaric imagination." As we all know, religion has not yet fully shed its darker side, and many honest critics think it never will. Is Whitehead wrong, however, when he adds that religion, in spite of its inevitable defects, "is our one ground for optimism," and that "apart from it, human life is a flash of occasional enjoyments lighting up a mass of pain and misery"?[3]

Both rightness and the religious awakening to rightness, in any case, mean something new when we interpret them in the context of a universe that is still coming into being. Over the past two centuries we have learned from science that the cosmos is on a long and far-from-finished journey—toward we know not where. Since it has already undergone many drastic changes over billions of years, we cannot begin to understand it unless we tell a story about these transitions. The remarkable thing about a story, I want to emphasize, is that it can be the carrier of meaning. We tell stories to bring out the significance of a series of events—for example, the episodes that make up our lives or those of our families, tribes, and nations. Over the past two centuries scientists, without setting out to do so, have been putting together a story about the universe. We cannot help wondering then whether this story carries a meaning too. Is it not conceivable that something of unquestionable importance is going on in this, the biggest story of all?

Or is the universe a story destined to end badly? As with any story, to answer such a question we have to wait. As long as a story is still unfolding, we cannot say exactly how it will turn out, and this is especially true of the cosmic story. Nevertheless, several significant stages of the story so far are clearly visible, and these may embolden our anticipation. Matter, life, and mind, three vast regions of being that used to seem like separate sheaves in a fixed vertical hier-archy, now show up as successive epochs in a still-unfolding drama.[4] This narrative drift of nature should be enough to make us wonder now whether something more is afoot. Can we be certain that wrong-ness will prevail? Dramatic arrangements and rearrangements of physical reality have already been happening, so who knows what may yet be taking shape on the future horizon of our story? At the very least we cannot help pondering what possibly comes next. Are our own lives, concerns, and struggles the end of it all—or perhaps the birth pangs of a whole new chapter in an even more dramatic unfurling?

Let us note also that stories may have distinct layers of meaning and that they can be read at different levels of depth. To find out "what is *really* going on" in any story we have to burrow beneath the surface, and not everybody is willing or automatically qualified to go there. Most of us are literalists, readers who remain content with shallow narrative impressions. Deep soundings of any story take training, discipline, and focus. In-depth understanding of what a story is all about does not usually come cheaply. Nor do we want it to, especially in the case of the cosmic story. Why should we lazily assume, after all, that this—the greatest story ever told—has no hid-den meaning, or that nothing important is going on deep beneath nature's narrative outer face? Is it self-evidently wise to declare, in the words of a contemporary American philosopher, that "the only message the universe has is that there is no message."[5] Is it not at least worth asking whether the universe carries a narrative filament too fine for cosmic literalists to feel? In all seriousness we still have reason to ask what is really going on inside the universe whose out-ward journey the natural sciences have barely begun to map.

In addition to cosmic literalists, of course, we are aware also of religious literalists whose cheap interpretations of their holy books are the source of untold mischief and misery in our world today. Literalist readings of ancient scriptures can give sacral, hence "eter-nal," legitimacy to racial, social, and economic injustice, even to massacres, suicide bombings, and genocides. All of this is part of what I will be calling the wrongness of religion. Yet the same sacred books that literalists interpret malevolently and anachronistically may also promote justice, social inclusiveness, ethical heroism, and undeniable sanctity.

We may read the new scientific story of the universe also at various levels of depth. Science has discovered that the universe story began with a "Big Bang" 13.8 billion years ago. To get a sense of the temporal immensity of this great epic, imagine that you have in your library a shelf containing thirty large books. Each volume is 450 pages long, and each page stands for one million years. The Big Bang takes place on page 1 of volume 1, and the first twenty-one books have to do only with lifeless physical, chemical, stellar, and galactic processes. Life is not in a hurry to make its entry into the story. Our solar system appears at the beginning of volume 21, about 4.5 billion years ago, but the earliest instances of life do not show up until volume 22 (3.8 billion years ago by many recent estimates). Life remains single-celled until toward the end of volume 29, where the Cambrian explosion occurs. At this point in the story organisms begin to become increasingly complex at a more accelerated pace than ever before. Even so, dinosaurs do not show up until around the middle of volume 30, and they go extinct on page 384. Only the last 66 pages or so of volume 30 feature the flourishing of mammalian life. Human-like ancestors begin to show up a few pages from the end of volume 30, but anatomically modern humans make their appearance only about halfway down page 450. Self-reflective subjectivity, ethical aspiration, and the religious quest for rightness arrive in the universe only in the last paragraph of the last page of the last volume.

After the first living cells appear late in volume 22, Darwinian evolution takes over, and with it comes a fixed formula for life's ongoing self-transformation. Evolution has three main components: random changes in natural history and heritable genetic patterns; the impersonal "law" of natural selection; and a vast amount of time. The cosmic process uses this evolutionary recipe to cook up all the countless forms of life, including our own species, thrusting us into a life-world that has been thriving quite well without us. No sharp biological break separates us from our animal ancestry. We are relatively recent products of a long and patient natural process that, from all that science can tell, did not have us in mind. We seem to be a cosmic afterthought with no reason to think of ourselves as special.

My wager here, however, is that the cosmic story has layers of meaning that remain hidden to casual readers and even to the scientific experts who are piecing the story together bit by bit. It is instructive, I believe, to look inside the story. The cosmos, we shall see, has hosted an inside story from its beginning billions of years before Homo sapiens came along to read it. So scientific accounts need to be balanced by a look beneath the surface, and this means, perhaps above all, searching for a new understanding of religion in the age of science.

Since science is interested in outward, measurable events and qualities, it passes over the inside story. It cannot be faulted, though, for reading nature in an impersonal, detached, mathematical way. Science's exclusively outside perspective is perfectly appropriate to its self-limiting method of inquiry. It rightly ignores religious concern about the meaning of it all. It is not part of good science, in other words, to look inside. However, neither is it part of good science to claim that there is no inside story at all. Such a declaration has no scientific standing whatsoever, even if it is mostly scientifically educated people who make it. In looking for a possible meaning to our universe, we cannot ignore scientific discoveries, of course, but we need to go beneath and beyond science in reading the whole story, both inside and outside. It may be, after all, that the point of the outside story is to carry an inside story. If so, a fully attentive reading cannot ignore the inner drama of a universe gradually awakening to the dawning of rightness.

Ways of Reading the Universe

I believe there are three main ways of reading the new cosmic story: archaeonomic, analogical, and anticipatory. Archaeonomy, my name for the first reading, assumes that we can understand what is going on in the universe only by digging back into its remotest past, excavating the series of physical causes that have led up to the present from the beginning. The word "archaeonomy" pulls together two common Greek terms: *arche*, meaning beginning or origin, and *nomos*, meaning law. An archaeonomic reading of the universe assumes that everything that happens in the history of nature is predetermined by inviolable physical laws established from the beginning.

I use the term "archaeonomy" to designate a worldview based on the intellectual assumption that the original or "archaic" phase of cosmic process "lays down the law" for everything that happens subsequently. I have sometimes called this first way of reading the cosmic story "archaeological," but here I want to avoid any confusion of archaeonomy with the splendid empirical scientific discipline of archaeology.[6] Journeying back into the cosmic past archaeologically is a necessary prelude to an inside reading of the cosmic story, and as such it is unobjectionable. In fact, understanding the universe requires our digging back into its past. For example, biologists are now looking for the origin of life. Their method of doing so includes breaking down currently available organisms and cells into their most elemental units. The expectation is that this analytical approach will help scientists reach an approximate understanding of the origin and earliest stages of life. In a similar way astrophysicists are trying to break down currently existing particles into finer and finer constituents and, in this way, reconstruct with the help of mathematics and high-energy accelerators the earliest stages of cosmic process.

Breaking things down analytically and exploring the past history of nature archaeologically, I want to emphasize, is essential to scientific understanding. When I use the term "archaeonomy," however, I am referring not to scientific analysis or archaeological exploration of the past but to a contemporary comprehensive metaphysics—a whole set of beliefs—that considers analytical or archaeological scientific inquiry the only right way to understand present phenomena. Archaeonomy assumes that the cosmos can become fully intelligible here and now by our recovering the elemental, subatomic simplicity of its earliest or archaic stages and then tracing a series of efficient and material causes forward from the remotest past into the present. The archaeonomic reading of nature, then, is "deterministic." It is based on the conviction that present and future states of nature are fully governed or lawfully determined by what happened in the remotest cosmic past. Archaeonomy also implies that since the laws of nature were firmly fixed at the very start of the cosmic process, and we humans are purely physical products of this process, our sense of human freedom has no basis in reality.[7]

The pure archaeonomic naturalist assumes that physical science's retrieval of the causal past is the only reliable avenue to understanding the universe, and that the world available to scientific analysis is really all that exists. Archaeonomy professes to be fully explanatory rather than modestly exploratory, whereas in fact it is a belief system and not a verifiable body of knowledge. Archaeonomy, as I use the term, is oblivious to the possibility that the cosmic story may have a dramatic interior meaning undetectable by science's outside, analytical reading. When I speak of materialism, naturalism, or scientific naturalism in this book, I am referring to the archaeonomic belief according to which elemental physical units alone are truly real, nature is really all there is, an "outside" reading of the story is sufficient to understand it, and any sense of a deeper meaning in cosmic process is illusory.

Even though an archaeological method of scientific inquiry has done well to make us aware of the main outlines of the new cosmic story, archaeonomy is a materialist belief system for which there is no scientific support. Since the "way of archaeonomy" explicitly rules out any inside readings of the universe, it thinks of religion as pure fiction, as a return to sleep rather than an awakening. The pure archaeonomic naturalist believes, but cannot demonstrate scientifically, that the universe is simply a series of physical states with no inherent or lasting importance. Archaeonomy therefore logically entails "cosmic pessimism." Cosmic pessimism—allegedly based on science—is a belief that the universe is altogether devoid of lasting meaning. If that belief is right, then religious symbols, ideas, and aspirations are nothing more than human concoctions.

Analogy is a second way of reading the natural world.[8] It looks upon the perishable things in nature as, at best, imperfect representations or analogies of eternal and invisible originals existing beyond the empirically available world. Following many centuries of otherworldly religious preoccupation, the "way of analogy" fixes its attention on what it takes to be the "eternal present," a realm of perfect being said to exist beyond the physical world of becoming and perishing. God is just one of the names religious devotees of analogy have given to this eternal present. Brahman, as mentioned in the Introduction, is another. Since the physical world is perishable, analogy is relatively indifferent to what goes on in nature independent of its sacramental significance for religion. Today devotees of analogy, especially perennialists, have almost no interest in looking for meaning in the long evolutionary story of life or, more generally, in the 13.8 billion years of cosmic process.[9]

According to analogy it is religion, not science, that awakens us to the real world. Religion's purpose is therefore to look beyond, or through, the perishable physical realm that science investigates. It seeks a perfection that lies above, beneath, or within the passing physical world. Analogy does not always oppose science, but it considers science of little or no interest spiritually speaking. In effect, then, because its attention is fixed on "another dimension," analogy no less than archaeonomy usually doubts that anything of lasting significance is really going on over the long haul in the physical universe.

Let me add here that the "way of analogy," which has deep roots in myths that long preceded the birth of philosophy and science, has also left an imprint on contemporary mathematical physics. "Analogical physics," not unlike analogical religion, is an increasingly popular body of speculation by scientists who, like many of their religious ancestors, deny that time is real. Analogical physics assumes, quite Platonically, that the real world is the timeless one depicted by mathematical understanding. Like analogical religion, its fixation on timelessness renders it oblivious to the possibility that something of significance may be working itself out in time and human history.[10]

Anticipation, a third way of reading the universe, eagerly embraces the new cosmic story. It considers time to be real and not an illusion. Anticipation allows that more-being or fuller-being can emerge in time.[11] Aware that the cosmic story is far from over, it looks patiently and expectantly ahead for a possible meaning to it all. Anticipation is especially grateful for scientific discoveries that invite us to understand the universe as an unfinished story. Unlike archaeonomy and analogy, the "way of anticipation" wagers that something significant is working itself out in the universe now as in the past. It reads the cosmic story both scientifically and religiously, from outside and inside simultaneously. It fully accepts evolutionary biology, Big Bang cosmology, and other scientific fields of research that contribute to our sense that nature has a narrative constitution.

Since stories can be read at different levels of depth, however, anticipation looks upon the new cosmic story not as an alternative to religious narratives, as archaeonomic naturalists do, but instead as an invitation to ask what is really going on in the universe, including in its recent religious awakening.

For anticipation the religious awakening to rightness is a cosmic and not just a human occurrence. Anticipation, as the label suggests, is willing to wait—not passively, but attentively and opportunistically—to see where the cosmic drama may be taking us and how we humans may contribute to it. Since the universe is still coming into being, the anticipatory vision does not expect anything to be fully intelligible at present. It does not insist on perfect clarity or absolute certitude here and now. It can live with ambiguity. Its dramatic reading of the cosmos tries not to possess but only to open itself to whatever meanings the story may be carrying. After all, by any standards of observation, there is no evidence that the story is anywhere near being over. Anticipation therefore looks inside the cosmic story only by simultaneously looking toward what may be dawning unpredictably ahead. Unlike archaeonomy, it does not insist, dogmatically and impatiently, that there is no point to it all. And unlike otherworldly forms of analogy, it looks for meaning not outside but inside the story.

While writing this book, I have become more convinced than ever before that analogy and archaeonomy, even though each is more dominant in the shaping of contemporary culture and consciousness than is anticipation, are ill-equipped to read the new cosmic story in depth. The analogical reading fails to appreciate sufficiently the reality of time and the fact that the cosmos is an ongoing story. Analogy is mainly a religious reading, but vestiges of analogy remain, as I noted earlier, in the speculations of contemporary cosmologists who ignore the historical nature of the universe and who regard as real only the mathematical principles that they can abstract from the concrete flow of natural process. The anticipatory reading that I am supporting, on the other hand, allows not only that time is real but also that more-being is coming into the universe from up ahead. It is also aware that the seemingly eternal laws of nature are themselves subject to change over the course of time, even though any such transformation may be indiscernible from within the limits of our own temporal and spatial situation.

For its part, archaeonomy's typically materialist reading explicitly denies that the cosmic process could be carrying a narrative meaning. In doing so, archaeonomy reveals itself to be not only a spiritual but also an intellectual dead end. Its fundamental incoherence, as we shall see, appears when it tries to account all by itself not only for the phenomenon of intelligent subjectivity but also for the real emergence of anything whatsoever.[12] I have listened for many years to boastful materialist claims that science unaided can potentially unlock all the mysteries of the universe, but it has not escaped my notice how little illumination materialist readings of nature have shed not only on religion but also on life, mind, morality, and other emergent phenomena.

I want to make it clear, once again, that in saying this I am in no way criticizing science itself, only those who expect too much of it. I am in awe of science, both its methods and its discoveries. I embrace Darwinian biology and the cumulative findings of physics, biochemistry, astronomy, and so on.[13] Nevertheless, while I find science illuminating, I am convinced that materialist interpretations of its discoveries only obscure their significance and dampen their intelligibility. The obvious fact of emergence—the arrival of unpredictable new organizational principles and patterns in nature—continues to elude human inquiry as long as it follows archaeonomic naturalism in reducing what is later-and-more in the cosmic process to what is earlier-and-simpler. A materialist reading of nature leads our minds back down the corridor of cosmic time to a state of original subatomic dispersal—that is, to a condition of physical de-coherence.[14] Unfortunately atomistic, archaeonomic materialism—a belief system that I distinguish clearly from science—persists as the conventional wisdom in much of the intellectual world today. It has also left its mark on what passes as Big History.

For centuries, impressive thinkers, obedient to the axiom that "something more cannot be fully accounted for by something less," sought to make life, mind, and spiritual existence intelligible by following the way of analogy. Analogy pictures nature as a static, vertical hierarchical arrangement of discontinuous levels of being. In the analogical reading of the universe life is more real and more important than matter, and mind more than life. Analogy accounts for the emergent moreness in nature by allowing for different degrees of participation in the eternal perfection of God. Analogy has appealed to religious people for centuries, but it remains intellectually plausible only so long as the universe is taken to be immobile. Once we realize that nature is a gradually unfolding narrative, we cannot help noticing that more is indeed coming into the story out of less over the course of time, and that it does so without miraculous interruptions and without disturbing invariant physical and chemical principles.

Analogy ignores, as religiously irrelevant, the scientific discovery that life and mind have come into the universe only incrementally over a period of billions of years. Archaeonomic naturalism, meanwhile, by reducing life and mind to lifeless and mindless material bits, offers no objective reason why we should appreciate life or mind as anything more, deep down, than precarious assemblies of physical units present from the beginning. If the analogical reading is unbelievable—since it has to bring in supernatural causes to explain how more-being gets into the natural world—the archaeonomic reading is even less believable since it cannot show how the mere passage of time accounts for the fuller-being that gradually emerges. If analogy cannot make the emergence of life and mind intelligible without bringing in a nonnatural mode of causation that lifts the whole mass up from above, archaeonomy is even less intellectually helpful in assuming that all true causes are ultimately mindless physical events, hence that life and mind are not really anything more than their inanimate constituents.

Anticipation offers a coherent alternative to both analogy and archaeonomy. It reads nature, life, mind, and religion as ways in which a whole universe is awakening to the coming of more-being on the horizon. It accepts both the new scientific narrative of gradual emergence and the sense that something ontologically richer and fuller is coming into the universe in the process. Instead of pretending to make complete sense of emergence archaeonomically, by looking back to the remote cosmic past, and instead of looking analogically for the "more" to drop into the world of "less" from up above, anticipation reads the universe as a story in which more-being and deeper meaning are always dawning on the horizon of the not-yet.[15] Not-yet, however, is not the same as nonbeing. It exists as a reservoir of possibilities that have yet to be actualized. It is a realm of being that has future as its very essence.[16]

"In nature every moment is new," says Ralph Waldo Emerson, expressing his own version of an anticipatory point of view. "The past is always swallowed and forgotten; the coming only is sacred."[17] Whereas archaeonomy leads to cosmic pessimism, and analogy entails an otherworldly optimism, our third option, anticipation, interprets the whole cosmic story as an ongoing awakening to a rightness that is both real and waiting to be realized. The cosmic awakening occurs successively in the emergence of matter, life, mind, and religion. An anticipatory reading of the cosmic story therefore requires a patient forbearance akin to the disposition we must have when reading any intriguing story. As we follow a story, its meaning at any present moment may be dawning, but it still lies mostly out of range. Reading the cosmic story calls for a similar kind of waiting, a policy of vigilance inseparable from what some religious traditions call faith. Indeed, there is a sense in which faith, as I use the term throughout this book, *is* patience.[18] Both archaeonomic cosmic pessimism and analogical otherworldly optimism, by comparison, are expressions of impatience. They share the belief that the universe must already have reached the fullness of being and complete intelligibility, whether in the remote historical past in the case of archaeonomy, or in an eternal present beyond time and history in the case of analogy. Each in its own way suppresses appreciation of the openly narrative character of the cosmos.

Archaeonomy, for example, after journeying back to the remote cosmic past, sweeps away the narrative whose trail science has uncovered. It interprets the long series of physical transformations that have occurred since the beginning as a pointless arranging and rearranging of primordial elements. In this reading there is nothing dramatic about the process. Unfortunately, analogy also virtually nullifies the drama of nature by treating the new scientific cosmic narrative as a mere curiosity at best. Even today analogical spirituality, along with analogical physics, continues to seek refuge from historical and cosmic time in a haven of timelessness. Like Indian traditions that identify ultimate reality (Brahman) as the fullness of being, my own tradition of Roman Catholicism nurtures spiritualities that are predominantly analogical rather than anticipatory. The main focus of these traditions has been on what is already eternally fixed rather than on what may yet be working itself out narratively in time.

Analogical religion on Earth, though deeply inspiring and consoling to human persons, is by nature unprepared to look for meaning in the still unfinished story of a temporal universe.[19] In saying this I wish not to disown or reject it but instead to situate its powerful sense of transcendent mystery in a new intellectual and historical setting. So I offer the "way of anticipation" as a third reading. Anticipation is attentive to both scientific discovery and the religious search for meaning. It appreciates the historical sciences—geology, evolutionary biology, and cosmology—that have laid open the narrative character of the cosmos. But the defining feature of anticipation is its expectation that meaning dawns only gradually, without magical leaps, as the story continues. Whereas archaeonomy cannot see more-being coming into the universe over the course of time, and analogy is not disposed to looking for anything significantly new as the universe continues its temporal journey, anticipation is willing to wait. Along with analogy, anticipation acknowledges rightness to be the greatest of present facts, but also, to repeat Whitehead's words, as something "waiting to be realized."

The religious awakening to rightness, as I have just proposed, demands a patience similar to the restrained attentiveness we must have when reading a complex story. In working our way through a great novel, for example, if we jump too far ahead, or stop reading out of frustration at not being able to make immediate sense of the disconnected items laid out in the opening chapters, we will surely miss the coherence that dawns only by our following the narrative watchfully all the way. So also with the cosmic story. We need to wait, not passively but with bridled expectation, for presently unforeseeable possibilities to make their appearance. Anticipation, in other words, enjoins its own brand of asceticism, not one of detachment from nature but one of patient, disciplined attentiveness to the possible arrival of fuller-being ahead.

An anticipatory approach reads the cosmic story in a manner that offers no necessary obstacles to either religious or scientific understanding. As I shall argue in more detail, anticipation assumes that when a new epoch emerges in cosmic history, it breaks no existing laws of nature while nonetheless changing everything dramatically. The universe's having an inner meaning, in other words, requires no suspending of the "laws" of nature as known by science. Allowing for the dawning of cosmic meaning requires no assumption on our part that the regulations operative in nature will have to be suspended at any point. The intelligibility that an anticipatory faith looks for in the universe is to be found not in scientific awareness of predictable physical patterns or organic "design," but by looking for a narrative coherence that gradually weaves itself into the recurrent patterns of physical activity as the cosmic process goes on.

In the real world, none of the three readings just summarized exists in a completely pure form. Hybrid versions are legion. Nevertheless, it is instructive to begin our study by making crisp distinctions. Briefly stated, the "way of archaeonomy" is vaguely aware that the cosmos is a story but finds no meaning in it. The "way of analogy" looks for meaning in the cosmos but fails to notice the story. The "way of anticipation" sees the story but insists that an in-depth understanding of it requires our waiting for its meaning to emerge.

Reading Religion

Corresponding to the three main ways of reading the universe, I see three parallel ways of reading religion. Today most academic interpretations of religion explore its origin and persistence from different scientific and historical points of view. These explorations have been illuminating, and I take advantage of them in this study. Nevertheless, in an unfinished universe nothing, including religion, can be settled simply by excavating its natural and historical roots. The problem lies not with archaeological exploration but with archaeonomic explanation, the assumption that uncovering a thing's past or its atomic makeup amounts to an adequate understanding of it. Scientific exploration helpfully discloses the story of nature and religion, but if the universe is still on the move, we can begin to understand it only by simultaneously turning our inquiry in the direction of what is coming. A posture of narrative anticipation is indispensable to making sense not only of the emergent phenomena of life and mind, but also of religion.

Archaeonomy, on the other hand, assumes that the key to understanding all current and future emergent phenomena, including religion, is to uncover their pasts. It is no accident, therefore, that archaeonomic assumptions govern most versions of Big History, including its understanding of religion. Archaeonomic naturalists take for granted that if religion can be explained in purely scientific terms, no room remains for alternative ways of understanding it. In reaction to the archaeonomic debunking of religion, contemporary devotees of analogy fear that the more tightly religion gets tied into physical, chemical, biological, or cosmological processes, the more likely it is that it will lose contact with the eternal present that alone can liberate humans from the travails of temporal existence.[20]

Neither archaeonomy nor analogy, as I observe throughout this book, befits the universe that science is now uncovering. And neither is prepared to understand religion in a manner that acknowledges its cosmic setting and significance. Anticipation, by contrast, reads both the universe and religion in one sweeping expectant vision. It accepts the scientific evidence that the universe is a long story still going on, but it rejects the archaeonomic assumption that nature and religion can become fully intelligible simply by our digging back into the elemental simplicity of their physical origins. Rather, anticipation wagers that in an awakening universe intelligibility lies ahead in unpredictable new narrative patterns and emergent meanings still out of sight. The intelligibility resident in still-emerging narrative patterns, then, can be approached at present only by our assuming a posture of patient attentiveness, scanning both the universe and religion simultaneously in a wide-ranging and long-suffering stance of expectation.

Anticipation endorses analogy's intuition that there is more to the universe than physical science can unearth. It allows that analogy is on to something of supreme importance. Unlike analogy, however, anticipation does not skip over the reality of time and cosmic becoming as though these were barriers on the way to right understanding. Instead, it reads religion as a new stage in a gradual cosmic awakening. It looks toward a universal religious meaning arising obscurely on the future horizon of cosmic becoming. To anticipation, the meaning to which religion points enters in not vertically from outside the universe but horizontally from up ahead. Meaning dawns on the horizon of the world's future in a way that draws religious attention ever more deeply into time and the ongoing cosmic story rather than away from them.

The "evidence" for a rightness that is always dawning, however, does not fall within the range of objects that science or ordinary experience can grasp and control. The dawning of rightness to which religious traditions refer by such terms as "revelation" or "enlightenment" can be registered only by adopting an interior posture of expectancy. True religious awakening, as we shall see, does not occur automatically. According to anticipation, what is most real arrives silently and unobtrusively, and we can greet it only by an adventurous leap toward the not-yet. By contrast, archaeonomy, safely fixing its gaze on the settled past, and turning its back to the dawn, remains officially blind to the rising of rightness. As ancient Daoism and other great traditions insist, however, what is most important usually goes unnoticed, so anticipation keeps turning eastward toward the gentle rising of rightness. The meaning dimly sensed on that front will seem real only to those who abandon a purely archaeonomic reading of the universe and wait patiently for rightness to arrive in a fresh awakening of hearts as well as minds.

The Road Ahead

I am reading religion here in the context of the universe's narrative journey through time. Religion in this perspective *is* the universe in a whole new era of awakening. In developing this cosmic understanding of religion, I do not offer a historical survey or comparative study of the world's religious traditions. Such works are plentiful, and I see no need to duplicate what other writers and teachers have done well enough already. My objective is quite different. As I said earlier, I want first to understand religion—especially in its axial search for a transcendent unifying principle of meaning, goodness, beauty, and truth—as an extension of the cosmic story. And second, I want to understand the new cosmic story from the point of view of the religious quest for an indestructible rightness. In carrying out this twofold task, I investigate twelve common religious themes to see what they might mean as part of a universe still awakening. In telling the story of the universe, contemporary scientifically informed thought has identified life and mind as distinctly new emergent phases that have arisen sequentially in cosmic history only after a long period of lifelessness and mindlessness. Religion, I want to add, is also a new dramatic stage in cosmic history. It has world-changing properties that only an anticipatory cosmic perspective can begin to detect.

Religion comes in many shades, but even so it has enough shared features everywhere on Earth that we may approach them collectively as a distinctly new emergent terrestrial and cosmic breakthrough. Religion is not just an aberration that a truly big history can cavalierly overlook. In each chapter of this book, I therefore single out a facet common to many religious traditions and examine its significance by placing it against the backdrop of an emerging cosmos. I have chosen for our inquiry general themes that are more or less visible in most religious traditions, but especially in those that Karl Jaspers associates with the axial period. The twelve properties investigated here happen also to be aspects of religion that scientific naturalists are striving now more than ever to demystify as childish illusions and/or evolutionary adaptations.

The featured topics are as follows:

1. dawning: a metaphor for the self-revelation of rightness—that is, of infinite being, meaning, goodness, truth, and beauty—that arouses the disposition of anticipation in religious experience and, by extension, in the universe itself;

2. awakening: the response not only of human persons and religious communities but of the whole universe to the dawning of rightness;

3. transformation: the dramatic process, both personal and cosmic, of becoming more, made possible by the world's ongoing awakening to the dawning of rightness;

4. interiority: the hidden, subjective dimension of the universe, life, and personal existence wherein the dawning of rightness is experientially registered;

5. indestructibility: the permanence of rightness, a quality that renders it immune to perishing and that delivers not only personal lives but the whole cosmic drama from the prospect of absolute death;

6. transcendence: the dimension of rightness that, according to axial religion, (a) extends infinitely beyond the reach of ordinary experience and science and (b) is an

endless source of new possibilities both for personal transformation and for the universe's ongoing dramatic awakening;

7. symbolism: the indirect, but highly privileged, mode of expression that religion must use in referring to transcendent, inaccessible, and incorruptible rightness;

8. obligation: the feeling of "oughtness" or moral duty awakened in personal beings by the dawning of rightness;

9. purpose: the aim or goal of cosmic process and personal lives that evokes the world's anticipation but is still in some sense "waiting to be realized";

10. wrongness: the undeniable fact that rightness has not yet been fully realized in the universe, in the life process, in human history, in personal lives, and in religion;

11. happiness: the expansive and peaceful feeling that accompanies the human anticipation of, and awakening to, the dawning of rightness; and

12. prayerfulness: the aspect of religion by which the universe explicitly awakens to indestructible rightness in human expressions of petition and gratitude, typically addressed to a personal deity.

My objective is to examine what each of these generic qualities of religion looks like when we place it in the context of our new scientifically anchored narrative of the universe. The list is not exhaustive, but all twelve items are aspects of religious awareness common to many traditions while also being highly contested in the age of science. Regarding the latter point, I intend to show that our awareness of an unfinished universe requires not only a new understanding of religion but also a new way of understanding religion's relation to science. The intellectual credibility—even the survival— of all religious traditions depends now on how convincingly they adapt their beliefs and aspirations to a scientifically understood universe whose spatial extension, temporal scale, and creative unfolding were unknown to religion's founders and main teachers.

As in previous works, I offer here an interpretation of religion that I consider to be completely consonant with the new scientific understanding of a universe in process. In doing so, however, I do not mean to imply that my interpretation of religion is superior to that of our prescientific ancestors. Their lack of a sense of deep time and their inability to acknowledge the unfinished state of the universe does not detract from the depth of their sense of the sacred. I still read the literature of my religious ancestors with the same degree of reverence and appreciation as before. My intention is not to replace their typically analogical quest for transcendence, but simply to ask what its sense of the infinite might mean in the context of a whole new understanding of the universe. Unlike the archaeonomic naturalists who take modern science to be a superior modern substitute for traditional religious awareness, I intend my anticipatory interpretation to bring together both the scientific quest for truth and the ancient religious sense of sacred mystery.

Finally, I want to acknowledge that my anticipatory reading of both religion and the cosmic story is inspired in great measure by tantalizing, but undeveloped, proposals in the writings of the geologist, paleontologist, and Jesuit priest Pierre Teilhard de Chardin (1881–1955). Teilhard was a deeply devout Christian whose religious life had been shaped by the analogical vision but who fell in love with evolution and the new sense of an unfinished universe. In my opinion he was also the first truly serious modern student of "big" history since he understood the need to look at the universe from the inside and not just the outside.[21] Teilhard achieved scientific acclaim in the first half of the twentieth century as one of the most important geologists of the Asian continent. So, quite clearly, he was able to look at nature from the outside perspective of the natural sciences. But he was convinced that a fully empirical attentiveness requires that we also look at the universe from the inside. Thus he could interpret the emergence of religious subjectivity as an integral part of the story of the universe, not just of human history. No modern or contemporary scientific thinker has tried harder to understand religion cosmically, and the cosmos religiously, than Teilhard.

As far as I know, Teilhard was the first scientist in the past century to have seen clearly that religion is an essential part of the inside cosmic story. I cannot ignore his original ideas. Although I am indebted to Teilhard's writings, however, I do not follow them slavishly. My integration of cosmology and religion is not identical to that of the great Jesuit scientist, so readers should not assume that this book is an exposition of his thought. Nevertheless, I agree with Teilhard that the thriving of authentic, anticipatory religion is indispensable to the future of the universe and not just to human persons and societies. "What is most vitally necessary" to the cosmic future, Teilhard said, "is a faith—and a great faith—and ever more faith. To know that we are not prisoners. To know that there is a way out, that there is air, and light, and love, somewhere, beyond the reach of all death. To know this, to know that it is neither an illusion nor a fairy tale. That, if we are not to perish smothered in the very stuff of our being, is what we must at all costs secure."[22]

CHAPTER THREE

Transformation

It seems to me immensely unlikely that mind is a mere
by-product of matter. For if my mental processes are determined wholly by the motions of atoms in my brain I have
no reason to suppose that my beliefs are true. They may
be sound chemically, but that does not make them sound
logically.

—J. B. S. HALDANE

RELIGION, AMONG OTHER TRAITS, has to do with transformation.
But so also does the universe. Fourteen billion years ago no thinking, knowing, or praying was going on anywhere in space and time
as far as we can tell. Nature was silent and unseeing. Now sight,
thought, and the quest for meaning have arrived, at least on Earth.
From aimless atoms the universe has undergone a transformation
into marvelous minds and restless hearts.

The cosmic journey from matter to mind may seem remarkable.
Doesn't science tell us, however, that it all happened so gradually
that we should not be too surprised? The coming of feeling and
awareness into the universe from out of an original state of insen-
tience has taken so long that there is nothing dramatic about it. Given

large intervals of time, ample space, and countless opportunities to experiment with different physical combinations, things can happen in nature over the long haul that would seem extraordinary only if they happened within a briefer span of time. The steps leading from mindlessness to mind have been so small, and the climb so gradual, that it scarcely seems uphill. And much backsliding has taken place along the way. The universe, moreover, may easily have remained mindless forever. Only a series of near misses let it become conscious and kept it from slumbering indefinitely. Additionally, a mindless "multiverse" may be loitering behind our own. If so, the phenomenon of "thought" that has appeared locally in our Big Bang cosmos seems all the more irrelevant in the total scheme of things. Consciousness has disturbed nature's normal state of sleep only momentarily, and it will eventually slip back into eternal forgetfulness.

Or so goes a fashionable archaeonomic version of the story. "Given a purely physical beginning," says the philosopher Simon Blackburn, "everything else—chemistry, biology, life, mind, consciousness, intelligence, values, understandings, even science—follows on by natural processes. Particles beget atoms beget molecules beget enzymes beget proteins beget life begets *Homo sapiens* who begets the Royal Society and the rules of tennis." Blackburn admits that "we do not understand every step in this process," but he has no doubt that science can close all the remaining gaps.[1]

Everything, including human subjectivity, according to this account, is the result of a drawn-out process of shuffling and reshuffling senseless atomic units. The dawning of human intelligent subjectivity seems unexceptional in this sequence of events. Everything adheres to the "principle of mediocrity" that has accompanied the rise of modern science from the beginning. In science's steady advance since the sixteenth century, things formerly considered remarkable turn out to be just average. First, Copernicus exiled Earth from its pivotal position in the solar system; second, our solar system lost its splendor after the discovery of billions of other stars in the Milky Way; third, our own galaxy shrank to a mere pinhead against the backdrop of two or three hundred billion other galaxies in the observable universe; fourth, life was shown to consist of lifeless chemical elements; fifth, our life-bearing Big Bang universe is now being swallowed up by an unconfirmed, but theoretically conceivable, multiverse. And all during this march of mediocrity the humanlike deities enshrined by popular religion for centuries on Earth have dwindled to the point of vanishing. It was almost inevitable, finally, that beings endowed with minds would also lose any semblance of being special.

Why Trust Our Minds?

The materialist picture of an unexceptional, mindless universe as narrated by Blackburn now seems intellectually secure—except for a glaring irony: The minds that think the universe is mindless do not act as though they are mediocre. Materialist philosophers do not write and speak as though their own mental prowess is reducible to mindless stuff. Are there any contemporary thinkers more confident of their cognitive abilities than renowned authors who glorify scientific materialism? How, then, can the swagger be justified if their materialist creed is true?

Let me put the question directly to Professor Blackburn: Given your archaeonomic doctrine of the "purely physical beginning" of the process that went into the making of your mind, how can you *justify* the exceptional confidence you have in that mind? How has your mind, which you tell me has its ultimate explanation in mindless past physical processes, become so ennobled as to be able now to arrive at right understanding?

Blackburn highly values his mind, sure of its capacity for discovering truths obscured throughout all past ages of prescientific ignorance. Yet if all outcomes of cosmic process, including minds like Blackburn's, are reducible to purely mindless material stuff, why should we pay any attention to them? Materialists tell us that all minds—and this would have to include their own—are products of blind physical processes and aimless evolution. Given such an ignoble lineage, if that is all there is to it, should we not pause before putting our trust in those minds? Charles Darwin asked a similar question: "With me the horrid doubt always arises whether the convictions of man's mind, which has been developed from the mind of the lower animals, are of any value or at all trustworthy. Would any one trust in the convictions of a monkey's mind, if there are any convictions in such a mind?"[2]

Refusal to ponder this paradox is almost the rule in intellectual circles today. As an illustration, along with Blackburn's narrative, we may cite the confident reflections of Perowne, a neurosurgeon in Ian McEwan's novel *Saturday*. Exulting in the archaeonomic account of nature, the protagonist croons without any sense of irony:

> What better creation myth? An unimaginable sweep of time, numberless generations spawning by infinitesimal steps complex living beauty out of inert matter, driven on by the blind furies of random mutation, natural selection and environmental change, with the tragedy of forms continually dying, and lately the wonder of minds emerging and with them morality, love, art, cities—and the unprecedented bonus of this story happening to be demonstrably true.[3]

This version of the epic of matter runs parallel to countless others today, most of them recited in a similar tone of unquestioning self-confidence. Yet it is hard to imagine a worldview more subversive of trust in the human mind than the archaeonomic materialism that inspires Blackburn and Perowne. The famous British scientist J. B. S. Haldane's observation that materialism is self-contradictory, as cited at the head of this chapter, seems irrefutable, and Darwin's doubt about whether, in the light of evolution, we are justified in trusting our mental activity merits serious consideration.[4] Yet the typical materialist evasion only digs a deeper hole. Given the enormous amount of cosmic time and the impersonal workings of Darwinian selection over several billion years, we are asked to believe, a mindless set of physical units can be transformed step by step into minds we can trust.

It is hard to avoid the suspicion that this claim is alchemical. The materialist version of mind's emergence maintains that the laws of physics and natural selection can transform the dross of mindlessness into the lustrous gold of justifiably self-confident minds simply by adding the elixir of deep time.[5] Certainly a transformation has been going on here, but to the materialist there is nothing dramatic about it. The blossoming of primordial cosmic mindlessness into organisms endowed with minds is nothing more at bottom than a patterning, scrambling, selecting, and repatterning of large numbers of blank bits of stuff across enormous spans of time. Yet the minds made by this process, we are to assume, deserve to be trusted anyway.

If mind is ultimately reducible to mindlessness, and the criterion of evolutionary success is reproductive fitness, how do I know that Blackburn's mind is not just adapting to the natural world rather than telling me the truth about it? But let me be clear. In no way do I contest physical and evolutionary narratives of the transformation of mindless matter into minds. I only question their explanatory adequacy. Blackburn is right to say that in the natural emergence of mind one step follows from another. But counting out how one thing follows from something else in any process is not enough to make it intelligible, as David Hume pointed out long ago. I accept the conventional, "outside" scientific narrative about life and the emergence of mind, and I enthusiastically embrace the new story of the universe, including evolutionary accounts of life and mind. Furthermore, I am open to evidence-based revisions of these accounts. What I question is not the gradualness of emergence nor the amount of time involved. Rather, what strikes me as objectionable is the archaeonomic downsizing of what is a truly dramatic cosmic transformation into a mundane sequence of physical states. By reducing a momentous transformation to mere transition, a purely physicalist account cannot satisfy the human drive to understand. The search for coherence—intelligibility, in other words—comes to a dead end whenever evolutionary accounts of life are conflated with materialist metaphysics.[6]

So it is not Darwin's narrative or Einstein's physics that I question. What I find troubling is the unscientific decree that an exclusively archaeonomic reading of the cosmic story is as deep as we can go in our understanding of it. I do not blame either Blackburn or McEwan's Perowne, by the way, for trusting their minds. It is right for them to do so, even impossible not to. What I am looking for is a way of reading the cosmic story that is coherent enough to justify that trust. Materialist literalism will not do the trick, as Blackburn's adversary Thomas Nagel has rightly pointed out.[7] Reducing our minds to mindlessness removes from the start any good reason for the spontaneous confidence we have in those minds.

No doubt, scientific materialists rightly trust their minds. But why? One reply is that social and cultural processes mold minds in a way that loads them with confidence. Since cultures are historically contingent and forever in flux, however, this proposal still washes out any secure footing for our cognitive pluck. How does one decide without logical circularity that a particular cultural setting is more successful than others in sculpting minds that can issue right judgments about the world? Is modern scientific method uniquely privileged in the history of human consciousness to justify our cognitive confidence? If so, how would we gather the scientific evidence to support that audacious claim? What warrants, then, can Blackburn, Perowne, and other materialists give us for trusting in the exceptional standing of their own cognitive activity? It seems to me that archaeonomic excavations lead us back, over and over again, to an abyss of incoherence instead of forward toward real illumination.

An anticipatory reading of the universe, I argue, provides a reasonable justification of our cognitive confidence. It is because our minds—along with the whole universe—are already grasped by the dawning of rightness that we can trust them. To drive home what I mean by this claim, however, let me begin by addressing you, the reader, directly. Notice that you cannot help trusting *your* mind at this very moment. You may not agree with me at first, but if you are questioning what I am saying here, it can only be because you do trust your mind's capacity to be intelligent and critical. In being obedient to your mind's imperatives to be attentive, intelligent, and critical, you give evidence of trusting your own cognitive abilities. Otherwise you would not be spontaneously raising questions, perhaps impassioned ones, about what I am writing here.[8]

Furthermore, you are not wrong to embrace the evolutionary scientific story of how your mind came to be. What is wrong is for you to embrace a worldview that consistently removes any good reason for you to trust in your mind. This, in fact, is what archaeonomic naturalism asks you to do. It implicitly subverts your cognitive self-confidence by telling you, sometimes in so many words, that your mind is reducible to mindlessness, as Blackburn's and Perowne's accounts imply. You need to have some degree of trust in your mind even before starting your intellectual and scientific excavations, but the archaeonomic "explanation," by tracing your mental life back to a state of original mindlessness, gives you every reason *not* to trust your mind.

How, then, *can* you justify your cognitive trust? Appealing to common sense, educational formation, and enculturation as reasons for your trust only begs the question, since all of these are awash in the flux of time-conditioned sociocultural currents. They offer no firm cognitive purchase either. Perhaps you will respond that it is all a mystery, and that you will never know why you can trust your mind. I believe you can do better than that. Catch yourself in the act of seeking right understanding as you are reading this chapter at the present moment. It is not hard for you to notice that your own mind is already attracted, at least tacitly, to what I am calling rightness. To be specific, you are looking for right understanding, another name for which is truth. Perhaps until now you never noticed your natural attraction to rightness. This is because rightness is not intrusive. Rightness beckons you gently and humbly. It does not force itself on you. Your mind comes in touch with it only by anticipation, not by possession.

How so? The rightness you anticipate as the goal of your desire to know is comparable to a horizon. A horizon is the field of all the things you can grasp from a particular point of view.[9] If you sit in a chair and look out the window, your visual field is restricted to the objects you can observe from where you are sitting. Now, if you stand up and stick your head out the window, your field of vision expands. When your point of view changes, the whole horizon shifts. If you try to grasp the horizon itself, however, it keeps slipping out of your control.

What is true of a visual horizon is analogously true of rightness. You cannot grasp it, but it looms silently—and humbly—as the horizon of all your conscious activity. It quietly grasps you more than you grasp it. It transforms you without dishonoring you. Apart from that silent horizon of rightness your own interior life—intellectual, moral, and aesthetic—would never have come into existence and would have no consistency or reliability. A horizon cannot be seized, but it can be felt and experienced by way of anticipation. It is your anticipation of, and being grasped by, a horizon of rightness, I suggest, that justifies the spontaneous trust you have in your own mind. You are drawn, moreover, toward rightness not only in your intellectual experience but also in your moral and aesthetic aspirations. Whenever you have an inkling of the difference between good and bad, or between elegance and ugliness, it is because you are quietly anticipating an elusive horizon of rightness.

The point of this brief exercise is only to convey that the quiet dawning of rightness now awakening and transforming the universe is not different from the rightness anticipated by your own mind—and most explicitly by religion. Religion is the trustful acknowledgment that the horizon of rightness is real and not something you made up. If you thought seriously that rightness is only a human or evolutionary invention, then your cognitive trust would have no justification. Only if rightness is real would you have a truly sound basis for trusting your mind.

Religion is a grateful acknowledgment that the rightness that quietly activates your mind, as well as your moral concern and aesthetic interests, is real even if it is incomprehensible. Acknowledging the reality of rightness, however, requires a personal transformation and a shift of horizons that materialists like Blackburn and Perowne are unable or unwilling to undergo. For them and countless other intellectuals today archaeonomic inquiry is enough, even though it leads our minds back eventually to the primal cosmic state of physical de-coherence. Hardcore scientific materialists are supremely confident of the rightness of their way of reading nature, and nothing seems to shake the spontaneous trust they have in their ability to reach right understanding. Logically speaking, however, the archaeonomic materialist worldview they take for granted cannot support their cognitive confidence.

Anticipation, Religion, and the Universe

A more internally consistent reading of the cosmic story might go as follows. After laboring long to produce life and mind, the universe did not cease waking up. Thousands of years ago it embarked on a process of transformation that eventually gave rise to religion, and along with it a sense of the reality of rightness. The vast majority of our human ancestors have been religious, so our specific biological name might well be *Homo religiosus* rather than Homo sapiens. In any case, religion is more than a biologically adaptive curiosity. It is a cosmic breakthrough, a new chapter in the story of an awakening universe responding to the elusive, unifying, transformative horizon of rightness. Since rightness is an ungraspable horizon, and not an object that can be grasped, there can be no scientific "evidence" for its existence. Its reality can be known only in the act of allowing ourselves to be carried away by it. And surrendering to it requires a deeply personal transformation that not everyone is eager to undergo.

Even so, it is hard to find in human history a religious culture that fails to require personal conversion as a condition of contact with what is really real. Likewise, it is hard to find instances of religion that lack a formal name for the unobtrusive rightness dawning on the horizon. "God" is one such name. Ancient Hebrews named it Yahweh. Jesus called it the kingdom of heaven or the reign of God. Hindus refer to it as *dharma* and *rta*, Daoists as the *dao*, Confucianists as *jen*, *li*, and *yi*, Sioux Indians as *wakan tanka*, ancient Egyptians as *ma'at*, Buddhists as *dharmakaya*. These are just a few of the many names Homo religiosus has bestowed on the horizon of rightness in whose ambience we may now locate the ongoing transformation of our unfinished universe.[10]

For much of recent human history the natural world has seemed but a stage, often a capricious one, on which the dramatic religious quest for rightness could take place. But in the setting of our new scientific cosmic story, religion is not just a human performance. It is the most significant way in which the universe announces its awakening to the horizon of rightness. A good reason for your cognitive confidence, therefore, is the common religious belief that rightness is real. If rightness were not real, you would have no solid justification for trusting your mind or assuming that the universe is intelligible. Unlike modern materialism, according to which rightness is a human invention and the universe inherently meaningless, religion insists that rightness is absolutely, unconditionally, and indestructibly real. The horizon of rightness does not go away no matter how hard you try to hide from it. Any serious criticism of religion, or any claims about the wrongness of specific religious ideas, can be carried out only if the critic too assumes the reality of rightness. It is the horizon of rightness that gives meaning to the universe and an incentive to human minds to keep exploring it. Through a wide variety of symbolic, ritualistic, cultural, and philosophical media, religion locates your mind and, along with it, the whole of nature within the embrace of a limitless, everlasting rightness. Encounter with this rightness, religious traditions claim, can bring you a sense of meaning, fulfillment, and lasting happiness. In doing so it can also justify trust in your mind.

I have said that we reach rightness, however, not by comprehension but by anticipation. To clarify what I mean by this claim I need now to contrast the anticipatory vision even more sharply than before with the more deeply entrenched archaeonomic and analogical ways of looking at the world. What follows is a necessary expansion on points I sketched only briefly in the previous chapter.

Archaeonomy. The archaeonomic worldview digs back into past natural history to make sense of present phenomena such as life, mind, and religion. Archaeonomy is the backward-looking and downward-drilling worldview that underlies Blackburn's and Perowne's readings of the cosmic story. It assumes that we can arrive at right understanding of everything only by scientifically excavating the series of physical causes that led up to them over a period of time. As the physicist Lee Smolin puts it, "If all that exists is part of nature, then all chains of causation must refer back only to past natural events."[11] Since we cannot actually go all the way back in time, however, archaeonomic naturalism takes a shortcut by attempting a fine-grained physical analysis of phenomena that are currently available. What science finds through increasingly minute chemical and physical analysis of living cells and other composite beings is nearly identical with what we would find if we could make our way back temporally to the remote cosmic past. For that reason, the archaeonomic vision may also be called the analytical vision of the world. Analysis is implicitly archaeonomic in its assumption that the most fundamental states of being are the preatomic stages prevalent at the time of cosmic beginnings. The purely archaeonomic or analytical reading of the universe, however, leads our minds not to fundamental but only to elemental aspects of the universe. Simplification is not the same as understanding.[12]

At the end of a scientific journey to the remotest period of the cosmic past we would come upon a vast, monotonous sea of scattered lifeless and mindless elemental units. Archaeonomy assumes that this region of past physical simplicity is the foundational state of all being. From its point of view, the only way to make life, mind, and religion intelligible now is, first, to reduce them to the lifeless and mindless physical elements from which everything started out; second, to construct mathematical models of how a combination of physical laws, random events, and natural selection has shaped all subsequent outcomes over an unimaginably long span of time; and third, to imagine how all future states of the universe can, in principle, be fully explained also as products of the same mindless elemental units undergoing manipulation by deterministic laws that have been in place since the far-distant physical past.

This archaeonomic reading therefore denies, in effect, that anything genuinely new can ever happen in the cosmic story. Real, dramatic transformation in cosmic history, in other words, is inconceivable. The main thing going on in the universe is a reconfiguring of mindless physical units across deep time. Present complexity, accordingly, is simply a "masquerade" covering up the simplicity of physical units and inviolable laws of nature. Everything going on in the universe is fully determined by the physical causes arising out of the remote past. So all present and future outcomes are hidebound to be nothing more than the uncoiling of what was already there.[13]

What is most distinctive and interesting about archaeonomic naturalism, then, is not its reductive, materialist, or atomist analysis of present phenomena. Rather, its most important feature is its making the chronological past the most fundamental state of the world's being. Pure archaeonomy is a distinctive worldview according to which the "really real" is what has happened already. I believe that archaeonomy's identifying what is fundamental and most intelligible with what has happened in the far distant past also quietly shapes the assumptions, interests, and methods of Big History.

While there can be no objection to recovering the cosmic past, the archaeonomic vision does not stop there. It disallows any room, in effect, for the arrival of more-being in present and future phases of the world-process. Archaeonomy, in other words, amounts to a "metaphysics of the past." Consider, for instance, the assurance with which the materialist philosopher David Papineau accounts for everything that ever happens in the universe in terms of temporally antecedent physical events:

> Physics, unlike the other special sciences, is complete, in the
> sense that all physical events are determined, or have their

chances determined, by prior physical events according to physical laws. In other words, we never need to look beyond the realm of the physical in order to identify a set of antecedents which fixes the chances of subsequent physical occurrence. A purely physical specification, plus physical laws, will always suffice to tell us what is physically going to happen, insofar as that can be foretold at all.[14]

The Duke University philosophy professor Alex Rosenberg, to give another of many current examples of the archaeonomic fixation, states: "Everything in the universe is made up of the stuff that physics tells us fills up space." Physics, he goes on, "can tell us how everything in the universe works, in principle and in practice, better than anything." In fact, "all the processes in the universe from atomic to bodily to mental, are purely physical processes involving fermions and bosons interacting with one another." Why, Rosenberg asks rhetorically, should we base everything on what particle physics has to say about the remote cosmic past? "Well, it's simple, really," he answers, again with unquestioned confidence in his own mind: "we trust science as the only way to acquire knowledge." And that, he concludes, "is why we are so confident about atheism." Intellectually speaking, in summary, archaeonomy places its trust in a worldview that locates the ground of all being in the remotest physical past, where, as we look back into it, the world fades off into subatomic dispersal—that is, into incoherence.[15]

Archaeonomy's metaphysics of the past assumes in principle that the world had already attained a fullness of being long ago in a physical state that analytical science alone can uncover in all its distributive rawness. Logically speaking, there is no room in this metaphysics for the universe ever to become more than what it has already been. Real cosmic transformation in the sense of bringing about something dramatically new and remarkable is therefore altogether impossible. Everything that has happened subsequent to the initial state of atomic dispersal is mere appearance, a cover-up for the blank plurality of physical units that lies beneath. This vision of reality, I must add, provides the philosophical basis for contemporary cosmic pessimism according to which the universe overall is inherently pointless, and religion cognitively empty.

Analogy. The analogical reading interprets the cosmos as a set of temporal imitations, or analogies, of a timeless supernatural realm of pure archetypes existing in heaven or in the mind of a supreme being. So we may also call it the archetypal vision. It looks upon worldly phenomena as imperfect copies of a primordial perfection hidden altogether from the archaeonomic point of view. A spirituality of contemplation is needed, therefore, to transform our minds and hearts so that they can read the universe as the revelation of an infinite transcendent rightness. In this spiritual setting all things in the temporal world are at best flawed imitations of a hidden eternal excellence. In some versions of analogy human thoughts, including those of religion, are dim remembrances of forgotten truths that had been revealed more splendidly either in the historical past or in a primordial revelation given from above.

The way of analogy includes Platonic, perennialist, Indian, and traditional Western theological worldviews in which nature is a limited, time-conditioned reflection of a heavenly excellence.[16] Assuming a "metaphysics of the eternal present," analogy implies that whatever gets accomplished in cosmic time and human history, except for the human soul's hidden growth in personal sanctity, is eventually nullified by being reabsorbed into an ever-present otherworldly completeness. According to analogy, the world process and human history exist only in the shadow of an "eternal now." I do not wish to play down the religious and ethical attractiveness this vision has for countless religious believers. However, in the final analysis, pure proponents of analogy have yet to show convincingly how cosmic time has any significance other than being necessary for the preparation of human souls for eventual communion with the timeless origin of all being. Analogy rightly understands human transformation as a process of becoming attuned to eternity. Yet it allows little or no room for the discovery by science that personal, human transformation is inseparable from a wider and deeper process of cosmic and evolutionary transformation. It pictures the cosmos itself as having been created complete in the beginning and then spoiled by the "Fall" of humanity from divine grace. It has no room for a universe that is becoming "more" over the course of time than what it was in the beginning.

I am arguing that analogical religion itself may now fruitfully undergo a process of transformation in the light of what we now know about the cosmos. Up until now analogy has been the main conceptual setting for interpreting the world's religious awakening to rightness. A major reason for its strong appeal even today is that it responds powerfully to human anxiety about perishing. Analogy offers an eternal refuge from the terrors of temporal and historical existence. It holds out a fixed haven from death and suffering. Not surprisingly it has little interest in secular achievements and technological innovations. Its many followers care little or nothing about evolution, the cumulative accomplishments of human history, and the larger epic of transformation that we call the universe. In contrast to the anticipatory vision, analogy idealizes detachment from the world as an essential condition for final union with rightness.

According to the analogical vision, you are permitted to trust your mind only because it "participates" (albeit imperfectly) in infinite being, goodness, truth, and beauty. You may have confidence in your native intelligence because it has never lost touch completely with the rightness from which it is now partly estranged. By teaching that everything here below shares imperfectly in what is timelessly true and perfect up above, therefore, analogy may justify the spontaneous trust you have not only in your mind but also in the universe's intelligibility. Analogy promotes a spirituality of contemplation designed to transform your mind and heart into ever more worthy receptacles for infinite truth, goodness, and beauty. Allowing yourself to undergo this conversion can make your life well worth living.

For the majority of religious believers even today, the way of analogy is the only plausible alternative to materialist interpretations of the world. The analogical reading of nature, whose roots reach back into preliterate human history, has been attractive to religious seekers for many centuries, and it is revered today all over the world in one form or another. Its promise to transform our minds and hearts makes a close study of the writings of its proponents—Plato, Augustine, Averroes, Maimonides, Aquinas, and Bonaventure, along with contemporary perennialists such as Huston Smith and Seyyed Hossein Nasr—a rewarding exercise. Its sincere quest for the infinite and for the unity of religious experience is worthy of serious contemplation. Nevertheless, analogy's unawareness of—and some-times indifference to—well-established scientific discoveries detracts from its claim to give an authentic reading of either nature or reli-gion. It rightly rejects the archaeonomic, materialist interpretation of nature as incoherent, but it wrongly ignores the full reality of time and the new scientific story of life and the universe. The up-shot is that by leaving out any close attention to the cosmic story it ends up with too narrow a notion of both rightness and religious transformation.

For the pure analogist, it cannot be true that the world is still being born. Or, if it is true, it cannot be important. Science's dis-covery that life and mind have emerged only gradually and that the world is aborning contradicts the prized analogical conviction that rightness is already fully realized. An untransformed analogical read-ing of the cosmos therefore cannot survive indefinitely in the age of science. Locating rightness in an eternal present above the flow of time, analogy turns its back on important facets of contemporary science and modern intellectual life. In my opinion, it cannot foster a full religious awakening as long as it fails to internalize one of the most sleep-disturbing discoveries in the history of human inquiry, namely, that the universe itself is a process of transformation still going on.

Anticipation. A reasonable alternative to both archaeonomy and analogy, I have been suggesting, is an anticipatory vision of the uni-verse. Anticipation approaches the world with a sense that rightness is still early in its rising and that its full dawning lies in a future be-yond present grasp. Rightness shows itself only dimly on a horizon that keeps retreating eastward the more we advance toward it and into it. We cannot seize it, but we can allow it to awaken and trans-form us—along with the whole universe.

Unlike analogy, therefore, anticipation is consonant not only with the new scientific story of the universe but also with traditional religion's intuition of indestructible being, meaning, truth, good-ness, and beauty—that is, rightness. And, in contrast to archaeonomy's attempts to reduce complex phenomena including life, mind, and religion to what is earlier-and-simpler in natural history, anticipa-tion keeps looking toward what is not-yet. It is ready to welcome the transformative influx of more being, meaning, truth, and beauty into each present moment. In doing so, it expects to encounter a rectitude that brings comfort to our anxious lives only because it simultaneously promises a new future for the whole universe.

To anticipation, no less than to analogy, the meaning of our lives has to do with our awakening to rightness. But in the case of anticipation meaning comes from opening ourselves, as part of an unfinished universe, to the full arrival of rightness from up ahead rather than just from up above. Anticipation respects analogy's insistence on the reality of rightness and its longing for communion with an ultimate, transcendent unity, but it also gratefully embraces evolutionary science and contemporary astrophysics for teaching us that the universe is an unfinished drama and for allowing us now to interpret religious conversion as part of a cosmic transformation. Anticipation cherishes analogy's mystical awakening to an infinite horizon, but it cannot turn away from the long cosmic itinerary to which our species and its aspirations are physically connected.

I believe the posture of active, attentive anticipation must now become fundamental to any religious vision that takes seriously the discoveries of modern science. Analogy seemed reasonable enough before we learned from science that the universe is still coming into being. Forgivably, religious people thought for centuries that the world had been created complete *in principio* or *en arche* (in the beginning). They could not have known that nature's earliest chronological state was one of elemental simplicity and that the cosmos was devoid of actual life and mind for billions of years. They did not realize that the still awakening universe may yet be open to fuller-being in the future. Science, however, has now so altered our understanding of the cosmos that henceforth any religious reflection that fails to take into account the unfinished state of nature can only be intellectually unacceptable and spiritually inadequate. It is the task of this book's remaining chapters to spell out more specifically what the ongoing transformation of the cosmos means for our understanding of common themes in religion everywhere.

CHAPTER FOUR

Interiority

The time has come for us to realize that to be satisfactory, any interpretation of the universe . . . must cover the inside as well as the outside of things—spirit as well as matter. True physics is that which will someday succeed in integrating the totality of the human being into a coherent representation of the world.

—PIERRE TEILHARD DE CHARDIN

IN THIS STUDY WE are no longer looking at religion simply as sentiments registered in the human psyche or as a peculiar kind of sociocultural activity that humans carry out as a way of adapting to nature. While it may be useful to look at religion in these narrower ways, here I am treating it as an unprecedented breakthrough in cosmic history. From the point of view of the elementary physical sciences, of course, nothing new or dramatic happens when religion comes along. No laws of nature are violated. Furthermore, the metabolic, neurological, and psychic processes that underlie all human longing are not suspended during the emergence of religious subjectivity. Yet with the arrival of religion something dramatically new is going on inside the cosmic story.

A conventional scientific approach cannot take us inside. The scientific study of nature is not wrong, it is just limited, especially when it comes to understanding the fact of subjectivity. Science can study religion as part of nature, but its objectifying approach tells us nothing about the actual content or truth-status of religious interiority. Scientific inquiry is neither expansive nor penetrating enough to put us fully in touch with the narrative quality of nature and the meaning of religion within that narrative. The natural sciences, in their search for understanding, strive to reduce complexity to simplicity, and uniqueness to universality. Things that seem remarkable from a dramatic perspective therefore look unremarkable from a purely physical, chemical, biological, neurological, or psychological point of view. When life made its first appearance in the universe 3.8 billion years ago, no physical laws of nature were broken, and even later, when nature gave rise to minds, there was no breakdown in the predictable routines of metabolism and natural selection. Physically, chemically, and biologically, everything continued as before. Yet from a cosmic/narrative perspective, the birth of life, and later thought, are developments that have brought something dramatically new and significant into being. So also has the arrival of religion.

The natural sciences are not wired to capture the drama of the universe's internal affairs. Physics and chemistry, for example, are content to resolve the objects of their study into chains of atomic units and impersonal material processes. Without interrupting the lawful regularities discoverable by physical and chemical analysis, however, life and mind introduced into nature a note of striving, drama, and pathos to which scientific specialties, because of their self-limiting methods of inquiry, are oblivious. The perspectives of physics, chemistry, biology, and other sciences do not have the breadth of vision or depth of feeling to register the turmoil and suspense going on in nature with the coming of life and mind into cosmic history. Even the dramatic entrance of conscious interiority into cosmic process caused not a ripple from the point of view of most scientific fields of inquiry. When the phenomenon of thought made its tentative debut in the natural world, no parades or trumpet blasts announced its arrival. And yet everything thenceforth was spectacularly different from before. Similarly, the arrival of religious aspiration has made a major difference not only in the story of human minds and social life but also in the epic of the universe.

No doubt, my claim that life, mind, and religion are chapters in a transformative adventure of cosmic awakening will find no intelligible place within the natural sciences, or for that matter in most academic disciplines. Furthermore, those who *believe*—I use the word deliberately—that analytical science alone can give us sufficient understanding and final truth will scoff at such a proposal. Any purely materialist reading of nature, however, is destined to leave out what is really going on inside. Instead of inquiring into the intelligibility of the cosmic story, archaeonomic naturalism keeps its gaze fixed firmly on the lifeless and mindless cosmic past. It closes itself off altogether to the inside story of nature by reducing the whole of reality to what can be specified only from the outside by fields of research such as physics, chemistry, and biology.

Archaeonomic naturalism, moreover, forbids our even asking what it means for a universe to awaken to rightness. Its typically materialist perspective downgrades to mere ordinariness everything we may initially have taken to be remarkable, including life, mind, and religion. From a purely physicalist point of view, the arrival of these three momentous epochs in cosmic history is simply the result of mixing and remixing elemental units across a vast period of time.[1] Moreover, since these emergent stages took so long to show up in cosmic history, materialists view them as purely incidental rather than as narrative keys to what the universe is all about. That the awakening of the universe has taken so long, however, does nothing to lessen the drama. Just the opposite: the slow passage of time, the enormity of space, the countless accidents and near misses along the way—all of these serve to heighten the narrative grandeur of a universe awakening from a long and dreamless sleep. The slow passage of time adds suspense and gives dramatic depth to the story. Along with the amplitude of space and the opportunity for innumerable experiments, the long temporal spread of the cosmos provides room for the playing out of an imposing story that we had known little about until recently. For all we know now, the story presently coming into view may be the carrier of a meaning that will always remain undetectable by purely scientific studies of the natural world.

Religion and Big History

As I am arguing in these chapters, only an anticipatory perspective can meet up with the inner drama of a cosmic awakening. Archaeonomy and analogy are not equipped even to notice, let alone appreciate, the inside story of a universe undergoing transformation into new modes of being and awareness. Analogy is aware of human interiority, but it cares little about the dramatic process that has gone into its production. In effect the analogical reading tears interiority out of its dramatic cosmic matrix and enshrines it in timeless heavenly splendor apart from nature. Meanwhile, archaeonomy—an approach that dominates contemporary academic inquiry—misses the inside story altogether. It connects human history physically to the longer scientific story of the universe without alluding to the interior commotion.

A fair illustration of this neglect is David Christian's *Maps of Time: An Introduction to Big History*. The book opens with several chapters nicely summarizing periods of natural history that readers may learn about just as easily by reading popularizations of scientific discoveries. Christian's book then follows this up with a sequence of chapters covering subject matter that conventional summaries of human history have made available already as well.[2] His project, I believe, is important, especially since modern Western cultures have tolerated for too long the ecologically dangerous assumption that human beings do not really belong to the universe. Linking our species to its natural prehistory, as Christian does, is also an educational necessity especially at a time when public denial of evolutionary science, and of human evolution in particular, is widespread. Like most authors associated with Big History, however, Christian fails to look beneath the outward flow of events to the momentous drama going on inside.[3]

The inside story of an awakening cosmos surfaces most clearly in the emergence of human consciousness, moral aspiration, longing for beauty, and religious restlessness. However, a dimension of insideness with at least some degree of thickness is coextensive with the whole story of nature.[4] A vein of interior awakening has long been throbbing in cosmic history, but conventional science is not equipped to take its pulse. Science is methodologically constrained to look only for physically specifiable connections. Its "outside" survey of the world, incidentally, also includes inquiry about dimensions of the universe that are invisible to us at present but whose physical existence can be confirmed in principle by scientific instruments of observation. Dark matter and dark energy, for example, are veiled from direct human vision, but they are not identical with what I am calling interiority since science can confirm their likely existence at least indirectly by way of its object-oriented and mathematical methods of investigation.

There exists a cosmic "inside story" of anticipation and transformation, however, that cannot be measured scientifically even in principle. It consists of occurrences that can be recorded only in centers of experience known as subjects. By the "inside story," then, I have been referring to all the events that occur in the hidden world of subjectivity. It includes sensations, moods, cognitions, desires, enjoyments, and—in the case of humans—moral and religious awareness, aesthetic sensitivity, and the longing for understanding and truth. This strand of cosmic insideness has gradually thickened and intensified since the origin of life less than four billion years ago, but a silent anticipation of the eventual emergence of sentient and conscious interiority has been part of cosmic process from the very beginning. The inside story coincides temporally with the outside story, and the meaning of the outside story consists somehow of its being a necessary condition for the existence of an inside story. Big History, however, typically fails to see the intimate connection between outside and inside.

Religion, in any case, is the continuation of a long process of cosmic awakening that began with the birth of the universe. By the time our human ancestors had begun migrating out of Africa more than 200,000 years ago, a stream of religious sensations and symbols had probably already been drifting from one generation to the next by way of linguistic, ritualistic, narrative, musical, and other early cultural vessels. Human genes, it goes without saying, are always factors in this transmission, but genetic "explanations" are still outside rather than inside accounts. No matter where, how, or when it began, there is an inside story of religious awakening that is as much a part of cosmic history as the formation of galaxies and the forging of the chemistry essential to life. Yet most versions of Big History, emulating science's focus on what is publicly accessible, avoid looking closely at the dramatic transformation taking place in the cosmic interior. "I assume the existence only of those entities that I can experience directly or can grasp using the methods of science," says Christian, Big History's chief spokesperson.[5] Indeed, most Big History publications, at least from everything I have been able to tell, assume a materialist—or what I am calling an archaeonomic—point of view, a way of reading the universe that permits only an outside perspective and looks solely for past physical causes to render present phenomena intelligible.

Taking for granted the intellectual superiority of secular culture and scientific modes of understanding, materialist historians such as Christian think of religion as a set of confused human constructs that have nothing substantial to tell us about the universe itself. They assume that only physicalist depictions are true, or at least approximately so, and that all religious myths about what might be going on in the natural world, including sacred stories of cosmic and human origins, are simply false.[6] The Harvard evolutionist E. O. Wilson, for example, takes all religious creation stories to be now obsolete "outside" attempts to understand cosmic origins.[7] Similarly, Daniel Dennett, another materialist interpreter of nature, reads the biblical creation stories literally and anachronistically as unsuccessful scientific accounts of cosmic and human origins that modern physics and biology have decisively put to rest. Seemingly unaware that the biblical authors could not possibly have had scientific interests, he smugly declares that Darwin's evolutionary account of life has rendered the biblical book of Genesis irrelevant.[8]

These archaeonomic readings of the natural world are identifiable not only by their scriptural literalism but also by their blindness to interiority as part of nature. This is why it is inconceivable to them that religion, like life and mind, may be an indispensable emergent development in a larger story of cosmic awakening. Nor can they contemplate the possibility that religious people arrive at meaning and truth in an entirely different way from the archaeonomic pursuit of outside, objectifiable "evidence." To scientific materialists there exists no transcendent realm of being that could conceivably be a factor in the genesis of religion, so they try to account for the latter solely in an archaeonomic way. Only science, especially evolutionary biology, they maintain, can say anything truly explanatory about religion's origin, symbolic content, and stubborn persistence.[9]

Materialist interpreters of religion, no doubt, are led to look at religion this way because they cannot fully acknowledge the reality of their own interiority. They generally have no awareness of the irony in reducing subjectivity—which, after all, includes their own cognitive activity—to utterly mindless material processes.[10] So whenever they encounter the "hard problem" of explaining subjective experience in terms of objective science, they either revert to dogmatic materialism, which cannot see subjectivity in the first place, or else they reluctantly admit intellectual defeat.[11] As they refuse to become fully aware of their own interior lives, it is not surprising that archaeonomic naturalists cannot entertain the possibility that the universe carries an inside story, one that has recently burst forth with unprecedented dramatic explosiveness in the form of religion.

Contemporary attempts by scientists and philosophers to unsnarl the knot of subjectivity are touching only the tip of a mountainous iceberg.[12] Beneath their efforts to make sense of mind in nature there lurks the deeper question of how to tell the whole cosmic story from both inside and outside simultaneously. I believe that only an earthquake in method and metaphysics can bring the inside and outside stories together into the richer synthesis needed to have a thickly layered big history.[13] Today the way of archaeonomy, unfortunately, continues to buffer intellectual culture from the distant rumbles of the needed new worldview. The most highly regarded studies of mind continue to interpret subjectivity as a fluke of natural history.[14] Subjectivity does not show up as something truly real on most current academically approved maps of the world. Yet there is no justifiable reason for leaving it off as though it were not an essential part of the cosmic story.[15]

Archaeonomy, moreover, fails to notice that the "inside story" of the cosmos includes not only animal and human awareness but also the whole set of physical processes that came into play in the early universe as an essential prelude to the dramatic flowering of subjectivity. The inside story, after all, did not make its appearance abruptly with the arrival of human awareness or even animal sentience. It started with the stellar, galactic, physical, and chemical transformations that prepared the way for the more recent arrival of explicit awareness. From a dramatic point of view, therefore, there could never have been a period in its history during which the cosmos was essentially mindless. The materialist thought-world, however, in its zeal to arrive at an exclusively archaeonomic understanding of the world, has virtually denied that subjectivity has real existence anywhere, let alone that it is a fundamental feature of nature as such.[16]

Western modernity will be identified in future—and bigger—historical accounts, I believe, as a brief chapter in the cosmic story during which the universe refused to face its inner side. Ours will stand out as an epoch in which reference to subjectivity became, in effect, taboo.[17] Late modern intellectual history will be looked upon as a dark episode during which the inwardness that accompanies life, consciousness, and religion was not only ignored but also at times taken to be nonexistent. In richer versions of big history yet to be written, wider-minded storytellers will point out that the expurgation of subjectivity and the removal of our own felt aliveness from the physical record gave rise for a brief time to what Hans Jonas calls an "ontology of death," according to which nonlife alone is given the status of true being. Our age will show up as a curious episode in cosmic history during which lifeless matter was taken, for a time, to be the "intelligible *par excellence*."[18]

A more circumspect big history will also record the calamitous ethical, political, and ecological effects of the modern erasure of subjectivity from authoritative portraits of the universe. It will observe that the expulsion of subjectivity sponsored by the current materialist worldview promoted the notion that animals are objects and not subjects, and it will express horror that the same point of view also tacitly sponsored the intellectual elimination of the human person from the sphere of being. It may add a footnote expressing puzzlement at how scientific materialists such as Steven Pinker, in the early twenty-first century, were cavalierly referring to human dignity as a "stupid idea."[19] They will be morally astounded that the theoretical elimination of subjectivity from the cosmos helped render millions of human persons vulnerable to the manipulative engineering and extermination projects that took place especially during the twentieth century.

Reductive archaeonomic physicalism removes from its picture of the world any reference to interiority, the only place where impressions of importance, value, and meaning could conceivably be recorded. Strictly speaking, scientific method appropriately leaves out any reference to subjectivity, and we can raise no reasonable objection to the use of such an abstractive kind of inquiry by scientists as such. Unfortunately, however, during the modern period science's methodological ignoring of inner experience has been utilized at times, against all reason, to support a worldview that formally denies that subjects exist at all. I do not have the space here to go into all of the ethical implications of this denial. Instead I must be content to observe that turning all of reality into a set of objects for scientific analysis and technological manipulation is not a morally innocent development in the history of human thought and the universe itself.

Especially troubling is the materialist "myth of objectivity," according to which nothing real exists outside the realm of quantifiable scientific comprehension. Such a diminished understanding of the scope of human cognition is incapable of acknowledging the reality, value, and importance of personal subjects. The myth's explicit denial of subjectivity has contributed at least indirectly, I believe, to the formation of intellectual and cultural beliefs that have in turn facilitated the mass killings of the twentieth century. It is still impossible for most of us to get our minds around the specter of many millions of people being slaughtered during this period as inconvenient objects standing in the way of the implementation of the economic and engineering visions of a handful of men such as Hitler, Stalin, Mao, and Pol Pot. It seems clear, however, that the myth of objectivity, especially over the past century, has made human persons seem like nonsubjects and hence vulnerable to being manipulated as mere objects. This monstrous chapter in human history cannot have occurred apart from intellectual conditions that had already drained the last strains of subjectivity from the universe. To be sure, one cannot blame the whole abysmal business of human slaughter and ecological wreckage on ideas alone. Yet the catastrophic dismantling of inconvenient social arrangements, the suppression of human rights, the outright murder of dissenting human beings, and the destruction of ecosystems essential to the survival of life can occur more easily in a world where subjectivity has disappeared from the map of the universe than in a world where it is acknowledged to be real.

Along with the loss of a sense of cosmic insideness, not coincidentally, comes an unprecedented assumption that the universe has no inherent meaning or purpose. The modern denial of cosmic purpose and the emergence of an unprecedented kind of cosmic pessimism, all in the name of science, are not without ethical consequences as well. There can be no objection, of course, to a scientific method of abstraction and quantitative analysis that brackets out any consideration of purpose. But archaeonomic materialism goes far beyond innocent methodological directives in its refusal to acknowledge that rightness is real, that subjectivity is part of nature, and that the cosmos may have meanings hidden from science. So before we are led astray by what may seem at first to be the elegant simplicity and commonsense realism of archaeonomic materialism, it is worthwhile asking whether this particular belief system is indubitably true or even approximately representative of the real world. I argue throughout this volume that it is neither.

Future and bigger historical accounts, I think, will applaud attempts by existentialist philosophers during the twentieth century to rescue human subjectivity, freedom, and personal dignity from the reigning materialist models of the world. They will wonder, however, why this well-intentioned rescue operation continued to tolerate the assumption that the nonhuman natural world is devoid of both subjectivity and hidden meaning. It seems to me, therefore, that a truly big, thick, and deep history has yet to be written. When such an account appears, it will be careful to highlight the irony of modern scientific materialists' denying ontological status to human interiority while at the same time tacitly trusting that their own intelligent subjectivity is somehow graced with an exceptional ability to discover meaning and truth![20]

A truly big—and thickly layered—historical narrative will look at the cosmos more generously and in a more anticipatory way. Rather than simply unearthing past temporal transitions in natural history, it will highlight the dramatic awakening of inwardness that has occurred in the evolution of sentience, consciousness, moral aspiration, aesthetic appreciation, and, above all, religion. Unlike modern objectivism, where the mere mention of inwardness is censored as intellectually evasive, a more penetrating historical inquiry will refuse to suppress wonder that the cosmos has always been in the process of awakening. It will be astounded that, during the modern period, intellectual life came under the spell of a worldview that associates the ground of all being with lifelessness and mindlessness. The more wide-eyed historical accounts we await will also list the names of a small minority of stubborn thinkers in the twentieth century who, because of their insistence on the reality of subjectivity, were ignored and even ridiculed by the scientifically elite at the time.[21]

Summing Up

Since the cosmic approach I am taking here will be unfamiliar to most readers, allow me to summarize and restate my argument up to this point. I have indicated that the virtual elimination of subjectivity from the cosmos by modern and contemporary thought renders most contemporary versions of Big History intolerably thin. Adherence to narrow archaeonomic assumptions keeps its proponents from attending to the insideness of things and to the anticipatory leaning of cosmic history that has reached its high point, at least so far, not in the arrival of mind but in the emergence of religion. From a biological point of view it may seem sufficient to observe that religion is an adaptation that facilitates the migration of populations of human genes from one generation to the next. Viewed cosmically, however, religion is a new chapter in an impressive adventure of awakening. From a cosmic point of view, religion is a relatively recent concentration of the anticipatory drift of a whole universe. The feeling of "being lost in the cosmos" that often accompanies religious subjectivity is not a signal that the cosmos is alien to us—as both analogy and archaeonomy take it to be. Instead, our religious uneasiness is an indication that the whole cosmos is still estranged from its final destiny. From anticipation's dramatic cosmic perspective, religion, in spite of what often seems to be its world-weariness, brings to conscious expression the incompleteness not just of our personal lives but also of an entire universe.

Archaeonomic naturalists, to be sure, will keep insisting that religion is not really an awakening but a way of remaining in the dark, or perhaps of returning to sleep. Isn't religion, they will ask, more like a shadowing of the mind, a barrier that shields us from the light of reason? Modern scientific naturalists claim that the universe began to wake up only with the arrival and maturation of science. According to modern materialists the cosmos is mindless and pointless, so religion can be nothing other than an escapist clouding of thought, an irrational flight from reality rather than a new epoch in the story of a cosmic awakening. As long as the universe is imagined to be essentially mindless, religious symbolism will seem, at least to the archaeonomic frame of mind—whether of Freudians, Durkheimians, or Darwinians—to be nothing more than an imaginative refusal to accept "reality."

Instead of looking at religion psychologically, sociologically, or biologically, however, we are now poised to look at it cosmically and in accordance with an anticipatory vision. After giving birth to sentience and the capacity to see, taste, touch, smell, and feel—traits common to both human and many nonhuman kinds of life—the cosmos recently gave rise to organisms not only capable of thought but also endowed with religious inclinations. To view religion simply as wishful thinking, as a sanction for social order, or as blind evolutionary adaptation would be to miss its dramatic cosmic significance. Even though archaeonomy may try to debunk religion by appealing to biology, neuroscience, and the social sciences, an anticipatory reading gives us another impression: The arrival of religion is the most dramatic episode of transformation ever to have taken place in cosmic history.

The new sense of a cosmic story brings with it an opportunity to broaden our understanding of what true awakening means. In modern times, before Darwin and Einstein altered our whole sense of nature, the cosmos more often than not seemed basically mindless and lifeless. The natural world was a neutral setting for life and human existence. It did not occur to scientists, philosophers, and theologians that the universe is a story still unfolding. People could not see that the birth of matter, life, mind, morality, aesthetic sensitivity, and religion are all part of a long transformative drama of awakening. To materialists even today the birth of mind, for example, is not a cosmic event but a localized fluke of nature that has occurred in the face of nature's overwhelming mindlessness. Evolutionary materialists have been telling us that the human mind is the result of a blind series of genetic accidents and adaptive experiments that have taken place relatively late in the cosmic process. To those of the archaeonomic persuasion, the phenomenon of "thought" is a momentary interruption of an enveloping cosmic silence.

Such an idea, however, as we have seen, is self-contradictory. Its exponents fail to notice how thoroughly their physicalist reading of nature, life, and mind sabotages trust in their own minds. Anticipation, on the other hand, by acknowledging that human minds are part of a universe already in the grasp of rightness, justifies the spontaneous trust we place in our native cognitive and critical capacities, and it backs up our almost automatic belief in the inherent intelligibility of the universe. To anticipation, the cosmos is a drama of awakening to the light of rightness, and religion is the most alert stage in that awakening. From the perspective of physics the cosmos looks like a process of heat exchanges and energy transformations, but from an anticipatory perspective the cosmos is an epic of gradually adapting to an ultimate environment of indestructible rightness. Subjectivity—the centered capacity to register experiences—has now reached a high point in cosmic history with the arrival of minds that can ask questions and awaken to meaning, goodness, beauty, and truth. In the recent birth of human consciousness the universe has taken an unprecedented dramatic turn. It has given rise, at least on Earth, to beings eager to understand where they came from, where they are going, and what they should be doing with their lives. And in the coming of religion the whole cosmos is now awakening worshipfully to the reality of a rightness that abides forever.

CHAPTER FIVE

Indestructibility

As for mortals, their days are like grass;

they flourish like a flower of the field;

for the wind passes over it, and it is gone,

and its place knows it no more.

But the steadfast love of the Lord is from everlasting to everlasting.

—PSALM 103:15–17

RELIGION IS A RESPONSE to the fact of perishing. It is the character-istic way in which human beings have sought to tie their fleeting lives to what is incorruptible. A refusal to accept death as the abrupt end of life shows up in the earliest human displays of animism and goes on today in the widespread religious longing for immortality, resurrection, and liberation from the cycle of rebirth. Religion has other commonly shared qualities too, and it comes in many flavors, but for nearly everybody its most intriguing, healing, and—in the age of science—uncertain attribute is its bold response to the fact of perishing.

What, though, does the religious longing for imperishability mean in terms of our new understanding of a dramatic universe? Sci-ence has now laid out in fine detail the narrative continuity of each human subject with the entire antecedent history of the universe. So from the point of view of both astrophysics and evolutionary biology the human self and the universe are a package deal. Every living organism, each perishable species, and each human person's life is an unrepeatable local flowering of the *whole* universe story. No single life can any longer be separated, whether by science, myth, or metaphysics, from earlier stages in the cosmic process. In the arrival of mind the universe began consciously to anticipate right understanding, but in the arrival of religion it clothed rightness in everlastingness. The cosmic significance of religion, then, is that in our personal acts of faith the whole drama of nature turns—in a conscious way—toward the horizon of indestructible rightness.

In the emergence of mind, whether this occurs only on Earth or abundantly throughout the universe, the cosmos has now become conscious of itself. Simultaneously it has also become aware of its capacity for nonbeing. Yet in religion human consciousness looks toward a final and decisive conquering of that abysmal threat. Starting as far back in human history as we can see, human subjectivity has been entertaining images, narratives, and (more recently) theologies that point to an imperishable state of being in whose ambience life overcomes death, light banishes darkness, and the cosmos finds everlasting fulfillment. In and through religion an awakening universe refuses to reconcile itself to absolute death.

Tying our lives to everlastingness has been, for the majority of its devotees, the most satisfying feature of religion worldwide and throughout human history. Even traditions that do not hope for conscious personal survival after death aspire to permanent release from the cycle of rebirth, and none of them pictures the universe as hurtling toward absolute nothingness. Even where there is no fully developed sense of subjective survival after death, as in the passage from Psalm 103 at the head of this chapter, we find a firm conviction that human life can be truly significant only if it stays connected in some way to an everlasting font of being. Instead of ignoring or debunking the bold religious intuition of indestructibility as mere adaptation, therefore, let us examine more closely what it may mean for our understanding of the universe after Darwin and Einstein.

If the arrival of life and mind made the universe dramatically different from what it had been before, the emergence of religion's intuition of indestructible rightness makes it even more so. Again, I am not denying that it is fruitful to look at religion psychologically, biologically, anthropologically, and in other ways. Yet we can gain a whole new perspective on the religious thirst for imperishability when we look at it cosmically. Religion, I believe, has cosmic significance simply by virtue of its intuition that an indestructible rightness lies beneath, above, beyond, or within "the passing flux of immediate things," as Whitehead puts it.[1] Through religion humans consciously connect the universe and every moment of its history to something inextinguishable.

From the perspective of cosmic history, religion is still in its infancy or adolescence, so it has room to grow. No doubt, up to this point religion has at times been mixed up with monstrous atrocity. In spite of all the moral objections we legitimately raise about religion's complicity in evil, however, we cannot rightly ignore its wide consensus that perishing is not final. And even though belief in personal immortality can be exploited as a factor in suicide bombings and terrorist violence, we need not for that reason alone assume that the religious sense of everlastingness is wrong. Moreover, religion's seemingly extravagant expectation of final deliverance from perishing need no longer be dismissed as childish escapism. A Freudian psychologist may claim that religious images of heaven are imaginative fictions rooted in the pleasure principle. And a Darwinian biologist may insist that religious intimations of immortality are functionally adaptive. Without contradicting psychological and evolutionary points of view altogether, however, we may also read religion's anticipation of imperishability as a vital part of the universe's awakening.

Before geology, biology, and cosmology revealed that our universe has a long past and in all probability an extensive future up ahead, it was analogy, in its many variations, that shaped human intuitions of a permanence beyond perishing. The study of religion, I believe, needs to acknowledge the importance of analogy in highlighting the human intuition of everlastingness. An awareness of permanence beyond perishing appeared in early religion, but it became most explicit in later analogical readings of nature such as we find in the Upanishads, Platonism, Zoroastrianism, perennialism, and classical Western theology. In Indian religion, for example, the empirically available world is sometimes thought of as a veil (*maya*) that covers up the real world of eternity. And in Western theology, following Plato, the temporal world is merely an imperfect reflection of a timeless realm of being in which the soul has its origin and destiny. Worldly existence, in that case, is a time of dreaming during which souls have only a dim remembrance of the eternal perfection from which they are now estranged. In a parallel way contemporary analogical physics privileges the timeless mathematical world to the point of regarding the passing temporal world as relatively unreal by comparison.

Analogy, whether religious or cosmological, is not concerned about the long-range future of the temporal universe. An anticipatory reading of nature, on the other hand, while also looking for indestructibility, refuses to seek release from the temporal world of nature. The meaning of time in that case is that it lets the cosmos exist in the mode of a story of awakening. Anticipation interprets the awakening to indestructible rightness as a property of the whole emerging universe, not just of individual souls. Instead of permitting personal souls to flee from time into eternity, anticipation waits for the whole universe to undergo an awakening and transformation. It hopes that when we personally die in seeming solitariness, instead of leaving the universe behind, our conscious subjectivity may, in a transformed state of being, enter more intimately than ever into the larger drama of cosmic awakening. If we are to be saved it is not apart from, but only along with, the universe.

The anticipatory vision allows also that our own limited lives and labors may contribute a modest but indelible new content to the ongoing cosmic story. Rightness, we have said with Whitehead, is not only a present fact, as analogy claims, but also something "waiting to be realized," as anticipation insists. Rightness transcends the universe, but it is vulnerable to undergoing internal change too by virtue of what happens in the cosmic story. From an anticipatory point of view, in other words, indestructible rightness is not lurking in fully finished splendor outside the universe, as analogy often assumes. Rightness is both a never-exhausted reservoir of new possibilities for cosmic becoming and a boundless compassion that allows nothing that ever happens in the universe to be lost absolutely.[2] The events that make up the outside story of the universe may perish, but on the inside there remains an accumulation of moments into a narrative coherence that is still both incomplete and beyond our grasp. The inevitable "heat death" of the temporal universe many billions of years from now therefore does not entail the permanent erasure of what has been going on inside the cosmic story. Even though the universe on the outside is made of perishing moments, the inside story is recorded and preserved everlastingly in the interiority of indestructible rightness itself.

The inside cosmic story consists of moments that instead of disappearing altogether as the story moves along, keep adding up. A defining feature of narrative, after all, is that its past moments do not dissolve into nothingness but instead gather into each new episode that follows. Otherwise we could tell no stories at all. Stories keep what has-been from vanishing altogether.

Narratives, in other words, have an irreversibility that can never be undone. After they are over, as the philosopher Charles Hartshorne points out, it remains forever true that they have taken place.[3] So even if the cosmos undergoes an eventual energy collapse, it will still be indestructibly true that the universe has existed and that it has undergone a unique and unrepeatable narrative journey. The inside story of the universe, a dramatic set of events that remains invisible to scientific objectification and mathematical representation, can never be wiped cleanly off the record of being. Moreover, if other "universes" exist, they too remain forever linked narratively to the whole set of events now transpiring in our own. They are not thrown away into an abyss of absolute nonbeing. Since it is the nature of story that earlier episodes and offshoots are preserved in its ongoing narration, anticipation trusts that everything that has occurred earlier in natural history continues to nourish each present moment throughout the whole process of cosmic transformation.

Archaeonomic Skepticism and Religious Anticipation

In different times and cultural circumstances religion has envisaged everlastingness in correspondingly diverse ways. Analogical religion renders that intuition most explicit with its notion of an indestructible realm of being to which each human soul belongs permanently. Anticipation, for its part, does not deny the fact of perishing, but it does deny the finality of perishing. Whether we read the transient, temporal world by way of analogy or anticipation, however, is not the religious sense of indestructibility an illusion that reasonable people must now dispose of in the age of science?

Archaeonomic naturalists, of course, unanimously insist that it is. The religious idea of a state of permanence beyond perishing, they say, is an invention of the human imagination. Yet even hard-core materialists cannot avoid postulating something everlasting on which to rest their own sense of the universe. Materialists not only appeal implicitly to the timelessness of truth whenever they look for right understanding, but they also steady their own lives and thoughts by tracing everything back to the immovable podium of the world's temporal past. In tracking everything back to a fixed cosmic past, the archaeonomic reading comes to roost on something that, in effect, will never go away. "One thing about the past. It's likely to last," says the verse maker Ogden Nash, and archaeonomic naturalists must agree.

Before modern science came along, many philosophers outside of biblical circles had assumed that matter is eternal and that mind-less physical stuff is the imperishable ground of all being. The ancient philosopher Democritus had claimed that everything that exists is reducible to atoms eternally arranging and rearranging themselves in a mindless void, and today archaeonomic naturalists, by embracing a strictly physicalist view of nature, cannot conceal their hunger for a similar kind of permanence beneath all perishing. In its attraction to something irremovable, pure materialism is not as distant from analogy as its proponents sometimes assume. The difference is that archaeonomy, contrary to analogy, assumes that subjectivity, if it has any real existence at all, is destined to be completely obliterated when the cosmos returns physically to the eternal sleep of original mind-lessness. Materialists usually deny that insideness is real at all, let alone imperishable. Mindless matter, they allow, may endure forever, but the awakening of subjectivity taking place now in cosmic history will prove in the end to be a lonely ripple in an oceanic oblivion.

Although archaeonomic naturalists these days do not consider the Big Bang universe imperishable, since the laws of physics are moving everything irreversibly to an eventual final catastrophe, nev-ertheless an increasing number of cosmologists are re-eternalizing matter in fashionable new ways. Having accepted the impermanence of our Big Bang universe, scientists are now smuggling permanence into their cosmologies through several different backdoors. One way is to assume the existence of a virtually imperishable multiverse, an indefinite number of unseen worlds lurking beneath, behind, or beyond our own. Another way is to suppose an endless series of cosmic oscillations—big bangs followed by "big crunches"—or a "mother universe" that gives birth endlessly to new offspring. Still another way of re-eternalizing nature is to suppose that a trail of "information" follows, however faintly, in the wake of all physical events. So even though things perish, the informational tracks they leave behind may never be completely erased. Finally, what I referred to earlier as analogical physics also assumes that eternity is real and not a human invention. It even imitates analogical religion both in denying that matter and time are real and in attributing indestructibility to a timeless sanctuary of pure numbers in which perishable beings may participate fleetingly and imperfectly.

In these examples we observe that the search for permanence in the midst of perishing remains a powerful dynamic underlying human consciousness, even in an age of alleged irreligion. Religion is the most explicit way by which people nowadays continue to deal with their anxiety in the face of possible nonbeing, but the quest for indestructibility is alive also in other guises in contemporary secular culture. I am referring here not to the vulgar quest to be immortalized for one's achievements, or the wish to have one's name mentioned for a few centuries after dying. Rather, I am thinking, first, of the increasingly enthusiastic popular as well as scientific interest in the prospect of extraterrestrial life (and intelligence) and, second, of the ecological longing by many people, skeptics as well as religious believers, to save the delicate tissue of life indefinitely. Significantly, astrobiology and ecology, in spite of the explicitly materialist beliefs of some of their most visible champions—for example E. O. Wilson and Carl Sagan—get their moral energy by feeding on a tacit longing to bring the whole of nature, and not just the human soul, into the range of redemption from absolute death. I am thinking also of how the archaeonomic vision itself is unconsciously driven by a mythic longing to overcome the erosions of time by getting back to the "archaic" timelessness of absolute beginnings.[4]

Unfortunately, a purely archaeonomic worldview, since it reduces life to lifelessness, cannot justify the implicit valuing of life

that undergirds contemporary astrobiological and ecological concern. Anticipation, on the other hand, can do so. Allowing that everything in nature is perishable, it does not treat life and subjectivity as
though they are destined for absolute nothingness. Nor does it understand living and thinking beings only as imperfect analogies based
on timeless archetypes now residing in a Platonic heaven or in the
mind of God. From an anticipatory perspective, things in nature,
especially living organisms, have value not because they participate
in an eternal present standing apart from time but because even in
their temporal passage they anticipate—and in this way become
advance installments of—the indestructible rightness dawning up
ahead. It is by anticipation, rather than analogical participation, that
the world here and now communes with indestructible rightness.

Analogy, by comparison, channels its interest in imperishability
toward a timeless heaven. Before the age of science there usually
seemed to be no other way of entertaining the ideal of immortality.
So religion today can be grateful to analogy for keeping alive for centuries our native attraction to the prospect of indestructibility. In
doing so, however, analogy has inevitably—though understandably—
ignored the net increase in being and value that has been going on
in the cosmic drama for billions of years. It is still inclined, therefore,
to regard nature mostly as a process of decay, and in most instances
it even considers all of time to be a deviation from eternal rightness.
Analogical otherworldly optimism thinks of the perishable world
as a setting from which the soul must seek release after proving its
moral and spiritual worth. For that very reason, it cannot appreciate
the narrative nuance and tension that make the cosmos a suspense-
filled dramatic emergence rather than a fatal decline.

Anticipation, no less than analogy, refuses to suppress awareness
of the world's imperfections and the fact of perishing, but it takes
them to be marks of an unfinished universe rather than unambiguous
evidence of a fall from primordial rectitude. Anticipation situates
each passing moment within a narrative accumulation of other passing moments. All moments in the cosmic story are passing, but they
can add up narratively to something both new and imperishable.
From the anticipatory point of view, indestructibility consists of a
fullness of time rather than of a changelessness beyond time. What
happens in time, then, does not simply disappear into an ontological

black hole. It builds up. Time, by keeping everything from happening at once, is a gift that allows the universe to be a drama. And drama can be the carrier of meaning.[5]

Anticipation, therefore, is even willing to follow archaeonomic journeys as they lead back to the universe's earliest, particulate state of being. Having arrived back at the remote preatomic past, however, anticipation then leaves the archaeonomic naturalists behind to sift the sand to which their analytical dig has led them. Instead of reclining complacently on the mound of elemental units into which materialist atomism dissolves the universe, anticipation adopts the pose of waiting patiently, though not passively, for new coherence to emerge farther along on the road of cosmic becoming. It rejects the archaeonomic assumption that nothing really new or "more" can emerge as the story unfolds. After all, as I have already pointed out, the materialist's own recently emergent mind must somehow be more than reshuffled primal particles if trust in its performance is to be justified.

In an anticipatory reading, the cosmos even at its birth was already turning—heliotropically, one might say—toward the dawning of more-being. Take yourself back in time to around 13.8 billion years ago when the universe was still in its preatomic infancy. Had you seen the universe at that time and in that state of being could you have predicted with any precision the cosmic and evolutionary outcomes now observable? Could you ever have dreamed that from such an inauspicious beginning eventually would come organisms that feel, think, aspire to goodness, and even pray? Only by turning, in a state of patient expectation, toward the horizon of what is "not-yet" could you have prepared your mind for what was to happen later on.

Anticipation cherishes indestructibility, but unlike analogy and archaeonomy, it does not associate imperishability with the fixity of the past. Instead, what is imperishable is the never exhausted dawning of more-being up ahead. Consequently, in the story-shaped universe we actually inhabit it is always wise to wait. To anticipation, waiting attentively for the story to develop, and for the coming of more-being, is epistemologically essential to right understanding—as well as to human happiness, as we shall see in Chapter 11. Anticipation does not expect anything to be fully intelligible yet—since the universe is still unfinished—but it finds sufficient satisfaction, at least for now, in its wayfaring expectation of a coherence not yet realized. And it is eager to engage in the kinds of human projects, including scientific research, that require patient anticipation of more-being. Analogy and archaeonomy, on the other hand, are easily paralyzed by epistemological impatience. They both demand complete clarity here and now. Not finding its longed-for lucidity anywhere but in the remote cosmic past, archaeonomic naturalism declares that the emergent cosmos is pointless, even if it has led (incidentally) to the production of minds. Meanwhile, analogy takes flight toward an eternal perfection allegedly residing altogether beyond the physical universe. Anticipation, for its part, stays on the ground, keeping itself connected to the cosmos as it waits for an indestructibility that is both real and yet to be realized.

Getting Inside

I have been arguing that in addition to scientific study, a wider kind of empirical survey of the universe is needed to make contact with the cosmic interior. Science and Big History to their credit have done well in bringing to light the outside narrative, but an exclusively scientific approach lacks the tools to represent what is going on inside. What is needed is a whole new metaphysics, a coherent view of reality, that is conceptually wide enough to embrace simultaneously (a) the inner world of awakening subjectivity, (b) the outer world of scientific discovery, and (c) the horizon of indestructible rightness to which the cosmos is awakening. What I am proposing is an anticipatory vision of the world based on a "metaphysics of the future."

A time will come, of course, when our Big Bang universe can no longer sustain life, mind, and human communities. In spite of its awareness that the physical universe is sliding irreversibly down the slopes of entropy, however, anticipation is confident that what goes on inside the story can be preserved and even enhanced in the interiority of indestructible rightness itself. Archaeonomy, of course, assumes that the eventual energy collapse of the physical universe will spell the absolute end to everything. And analogy expects an immaterial aspect of each personal life to be rescued from the decaying cosmos by peeling off toward final immersion in an eternal present. Anticipation, however, relies instead on a vision in which each moment of the inside cosmic story is gathered up and pre-served everlastingly in a not yet fully realized narrative coherence.

I have suggested that you may embark on your own journey to the cosmic interior by first becoming aware of the anticipatory thrust of your own consciousness. While looking scientifically at the outside story, try simultaneously to become aware of your own awakening to being, meaning, truth, goodness, and beauty. In telling the inside story of the universe you cannot start at any clearly defined point of origin outside your own intelligent subjectivity. Nor can you pinpoint the exact moment when you first became a conscious subject. Searching for the outside origins of what is going on inside always fades off into uncertainty. You have no choice, therefore, but to dip into the stream of cosmic awakening by first becoming aware of your own interior cognitive life. You cannot circle around and grab the emergence of your subjectivity or the anticipatory thrust of the cosmos by the tail, as it were, since that effort would already be an exercise of your subjectivity, not a grasping of it.

Again, in getting to the inside story of the cosmos you have no better port of entry than your own inner cognitive life. While looking at how science links cosmos, mind, morality, and religion to one another, it is also right to pause and attend closely to what your own mind is doing right now. By all means, learn as much as you can about mind, morality, aesthetics, and religion from the various fields of scientific inquiry such as evolutionary biology and neuroscience. But science alone will never bring you over to the inside. No amount of objective scientific inquiry can ever tell you what it is like to be a sentient subject.[6] Likewise no accumulation of scientific information about cells, brains, and nervous tissue, or about the physical and biological processes that gave rise to them, can ever tell you what it is like to be a meaning-seeking and truth-oriented subject.

What you *can* learn from science is that the universe, from the beginning, has carried a set of physical features that have lent themselves to being patterned eventually into subjective states of being. By definition, however, subjectivity as such is concealed from objectifying consciousness. This is why archaeonomic naturalists, for whom all of being is reducible to objects determined by past physical causes, usually refuse to mention subjectivity, and even when they do, they typically judge it irrelevant or even nonexistent. Subjectivity, to materialists, seems to be little more than an illusion hovering insubstantially over elaborate arrangements of nervous tissue. Subjective states, archaeonomically understood, have only the thinnest claim to being. To anticipation, however, the emergence of conscious subjectivity brings a narrative coherence and intelligibility to what has transpired earlier in the cosmic story.

As long as subjectivity seems scarcely removed from nothingness, as it does to archaeonomic naturalists, its eventual dissolution would not cost the universe much, if anything. By leaving subjectivity, including their own, out of their portraits of the universe, however, materialists are asking us to embrace their materialist worldview as right. But on what grounds? What is the root of the rightness to which they appeal in their denial of subjectivity? What reasons can they give for accepting their own materialist understanding as right, and rejecting yours or mine as wrong, if neither subjectivity nor rightness is real? And if their own intelligent subjectivity cannot claim even the same degree of being as the mindless stuff from which it is said to have emerged, why should you or I pay any attention to them?

The story of mind-making, as contemporary astrophysics has shown in detail, has been going on from the time of cosmic beginnings. From its opening moments the physical makeup of the universe has been anticipating an awakening. Before giving birth to mental, moral, and religious subjectivity, the cosmos gave rise to life, and it was in the crucible of emerging life that subjectivity, latent in matter from the start, began to stir. At first inarticulate, subjectivity began to make itself present in acts of striving that are characteristic of all living organisms. We distinguish living beings from nonliving, after all, because we have an intuition that, like ourselves, all other living beings have to strive, to try, or to endeavor to achieve goals if they are to be and stay alive. But striving is impossible without centered subjects to do the striving. Whatever life may look like at the levels of physics and chemistry, from an anticipatory perspective the main mark of life is subjective striving. Every reader striving to make sense of this chapter may understand immediately what I mean.[7]

When life became self-aware, conscious human subjects began to strive not just for understanding but eventually also for right understanding, right action, and right satisfaction. It is in religion, how-

ever, that the anticipation operative in cosmic process from the outset makes its most explicit contact with indestructible rightness.
Intimations of an inextinguishable rightness are already implicit in
every person's anticipation of truth, goodness, and beauty. These
incorruptible "transcendental" goals are so interwoven with human
subjectivity that because of its association with them a total death
of consciousness is literally unthinkable. No doubt, anticipation is a
still small voice. It has to struggle mightily to make itself heard today
above analogy's otherworldliness and archaeonomy's covenant with
mindlessness. Anticipation, nonetheless, goes well with the contemporary scientific understanding of the cosmos as a continuing journey. It also falls into line with a religious interpretation of the universe as a still tentative, far from finished awakening to a rightness
that is forever.

CHAPTER SIX

Transcendence

Deeper than the soul of individuals, vaster than the human group, there is a vital fluid or spirit of things, there is some absolute, that draws us and yet lies hidden. If we are to see its features, to answer its call and understand its meaning, and if we are to learn to live more, we must *plunge boldly* into the vast current of things and see whither its flow is carrying us.

—PIERRE TEILHARD DE CHARDIN

I HAVE IDENTIFIED RELIGION cosmically as the grateful awakening of a whole universe to an indestructible rightness. I need to add now that religion generally thinks of rightness as extending infinitely beyond the physically available universe. Rightness then may never be grasped, but we may allow it to grasp us. Since we cannot comprehend or control it, religion speaks of rightness indirectly. It uses the language of symbol, metaphor, myth, and ritual. More powerfully than direct discourse, these modes of expression give religious people a palpable sense of being drawn toward an elusive horizon of all-encompassing being, meaning, truth, goodness, beauty, bliss, and (in many traditions) infinite compassion.

Religious thought refers to this horizon as transcendence. Sometimes transcendence means simply the process of "going beyond" present limits. At other times, however, transcendence means the infinite mystery of rightness that encompasses the universe and invites finite beings to enter into it ever more fully as a condition of their ultimate liberation and fulfillment. Western religion refers to it as God, but it goes by other names as well. The sense that rightness lies infinitely beyond ordinary human understanding is annoying to contemporary naturalists who decree that the totality of being should fall within the ambit of potential scientific mastery. For religious people, however, it is a matter of absolute importance that transcendence is not subject to human cognitive control. Their sense of an inexhaustible dimension of transcendence widens their world immeasurably. The encounter with transcendence brings unrestricted breathing room and, along with it, the prospect of arriving at a peace surpassing all understanding.

As the sense of transcendent mystery has become more accentuated ever since the emergence of axial religion, so has the intuition of its unifying power. Centuries of religious ferment have led to a widely shared mystical sense on the part of religious people all over the world that an unseen power of attraction gathers everything into a unity beyond the fragments of past and present experience. The powerful appeal of monotheism—belief in one God—lies in its promise of bringing the world's physical plurality and scattered events into a saving harmony not yet fully realized. In some cases, the divine creative synthesis is thought of as dissolving all differences, but in others, especially in the Abrahamic traditions, God is a power that unifies without abolishing differences.

Analogical religion maintains that the ultimate unity of all things is already a fact and that our impression of present plurality is illusory. Anticipatory religion, on the other hand, acknowledges that the process of unification is incomplete at present and that the meaning of our lives is to participate, with patience and forbearance, in the process whereby cosmic multiplicity is gathered gradually into a not yet fully actualized unity. Anticipation looks toward a climactic state of unity in which differences will be accentuated rather than dissolved. The archaeonomic reading of the cosmos, by contrast, forecasts the eventual victory of featureless plurality over differentiated unity. It predicts a final cosmic state in which atomic multiplicity and physical homogeneity will melt away all differences achieved in time. At that point the fragile physical syntheses the cosmic process has laboriously brought into existence will lapse back into a state of material dispersal in which life, mind, beauty, experience, and love will have disappeared forever.

An analogical reading of the universe gives primacy to unity over plurality, but the unity to which it directs religious attention is one that exists eternally, untouched by time. Redemption in that case means the individual soul's final communion with the timeless unity from which its sojourn in the disintegrating physical universe is only a temporary departure. The anticipatory perspective, like that of analogy, expects a final victory of unity over plurality, but, unlike analogy, it assumes that the unity, though real and present, is also in some sense "waiting to be realized." In an unfinished universe hoping that our own lives may contribute significant difference to the ultimate realizing of rightness can bestow on them a narrative and moral significance that neither analogy nor archaeonomy can allow.

I shall take the liberty of referring to the axial religious intuition of an infinite, transcendent unity as God-consciousness. Of course, not all traditions fit neatly into this recent development in cosmic and religious history. Buddhism and Daoism, for example, are not explicitly interested in promoting God-consciousness. Nevertheless, they are integral to the larger story of the universe's religious awakening. They seek "to transcend" the limits of ordinary human life and its everyday preoccupations. Their emphasis on the importance of true enlightenment is not easily separable from the general awakening of the universe to God-consciousness. Nontheistic traditions aspire in different ways to a state of being and awareness that lies beyond the general human experience of unsatisfactoriness. Their longing to transcend plurality, mediocrity, and mundane existence helps explain their appeal even to devotees of traditions centered on the notion of divine unity. Their refusal to reduce rightness to psychic projection protects them from easy association with contemporary science-based atheism. Above all, however, their silence about God points rightly to the inadequacy of all religious symbols of transcendence and to the poverty of every theological proposition.

The religious awakening of the universe to transcendent rightness is not only much wider but also considerably older than the

God-consciousness that emerged during the axial period. The inside story of the universe includes all the ages of awakening that have come to expression in religious subjectivity. I believe, nevertheless, that the axial emergence of God-consciousness is a decisively important breakthrough in the long religious awakening of the cosmos. The idea of God has cosmic significance primarily for its being a third major step in the drawn-out drama of cosmic self-unification.

The universe had undergone two major self-unifying transitions earlier. The first of these occurred when life emerged, at least on Earth, less than four billion years ago. During this creative period of cosmic transformation, previously scattered physical and chemical entities came together in remarkable new organic syntheses that we now refer to as being alive. And even though the laws of physics and chemistry underwent no violation when living organisms came onto the cosmic scene, with the arrival of life the universe departed decisively from earlier, simpler, and more predictable routines. No matter how geographically isolated the initial instance (or instances) of life may have been from the rest of the universe, the arrival of the first living cell marked a sharp new turn in the cosmic story. The novelty does not show up at the levels of purely physical and chemical analysis, but from a dramatic point of view the birth of life introduced an unprecedented synthetic note of striving into the cosmic symphony.

The universe has never been the same since. Even if it turns out that life has arisen on only one tiny planetary outpost (which seems unlikely), the entire universe, narratively speaking, has nevertheless achieved a remarkable new overall unity by virtue of this single local passage. Furthermore, if beyond our Big Bang universe there swirls a multiverse whose statistical immensity is necessary to increase the probability of life showing up at least somewhere, even that unfathomable lifeless totality of worlds nonetheless is given a new overall unity by virtue of its being connected in a dramatic manner to the arrival of life here on Earth.

A second momentous, and even more unifying, cosmic transition occurred when self-conscious minds—endowed with a passion for intellectual and narrative coherence—appeared on planet Earth. The emergence of reflective thought in the human species now allows the universe to become conscious of itself. Neuroscientists today are aware, of course, that human brains have antecedents in prehuman evolution, but the cerebral and neurological continuity we have with our animal ancestry does not rule out the fact that our power of thought brings the entire cosmos into a whole new epoch of interior self-synthesis. Certainly we share genetic and cerebral architecture with other animals, but with the arrival of humans in evolution there is a crisp new breakthrough in the story of the universe. As the neuroscientist Michael Gazzaniga notes, "While we can use lathes to mill fine jewelry, and chimpanzees can use stones to crack open nuts, the differences are light years apart. And while the family dog may appear empathetic, no pet understands the difference between sorrow and pity." In the arrival of human consciousness a whole new "phase shift" has occurred in the history of nature.[1]

I believe the novelty becomes especially visible when we consider the power of the human mind to bring the whole universe together from the inside. Only a willful suppression of wonder could lead the pure archaeonomic naturalist to ignore the drama of what is going on when the human mind arrives on the scene with its exceptional passion to discover an internal unity in nature and human history. Even though mind's emergence has been so gradual as to seem unremarkable from a purely physicalist perspective, an anticipatory reading of the universe cannot help observing that the appearance of minds obsessed with finding intellectual and narrative coherence amounts to a radical new transition in the drama of cosmic awakening. In the arrival of thought on Earth the cosmos dramatically folds back on itself and begins gathering into potential comprehensive unity every event that has ever happened in its long history.

The universe's passion for cognitive self-unification emerged early on in the human urge to tell stories. Myths about cosmic origins and destiny, for example, satisfy a very human longing to gather up the diversity of tribal and personal experience into an overarching narrative unity that gives meaning to life in common with others. Discovering the ultimate unity of all things is one aim of humanity's major creation myths. Modern science also exemplifies the unifying passion of the human mind. We measure the success of science, after all, in terms of its capacity to discover increasingly more comprehensive and unifying points of view. Science allows our minds to unify mathematically all the diverse patterns of nature. A longing for coherence even helps explain the intellectual appeal of archaeonomic readings that reduce the whole cosmos to the physical simplicity of its remotest past stages. Finally, Big History is a good illustration of how the universe is now seeking, through the instrumentality of human minds, a comprehensive self-unification across the long reaches of time, a synthesis from which nothing is supposed to be left out.

Unfortunately, in taking an exclusively outside survey, as we have seen, Big History leaves out the inside story. It settles prematurely for a unity that fails to include the dramatic dimension of interiority from which the human longing for narrative coherence springs. Focusing only on the predictable physical habits of nature, all archaeonomic readings of nature fail to appreciate the sensational dramatic transition going on when mind arrives in the universe's interior precincts. Like other ways of reading the world, archaeonomic naturalism cannot avoid looking for unity amid the world's plurality, but it does so by reducing everything including subjectivity to the monotonous physical uniformity prevalent before the emergence of life and mind.

An anticipatory reading, on the other hand, notes that mind's quiet "inside" entry into nature makes the universe spectacularly new once again. From an outside point of view the inviolable laws of nature still remain in place when mind emerges, but they function like grammatical rules rather than physical forces. When we humans tell stories, the rules of grammar remain consistently operative while new dramatic meaning is emerging during the narration. The syntactical regulations neither break nor bend in the process of storytelling. Nevertheless, the rigidity of the grammatical regulations does not prevent new and unpredictable content from slipping into the story. On the contrary, in the telling of a story grammatical strictness supports rather than inhibits the expression of new meaning.

Staring only at the inviolable grammatical rules, however, would divert our attention from the new levels of meaning emerging in any story. Analogously, a purely archaeonomic reading of nature distracts our minds from what is really going on in the cosmic drama. Materialist readers of the cosmic story get so fixated on the elemental components and "laws" of nature that they fail to notice the new narrative content emerging inside the cosmos, especially when nature becomes alive and eventually conscious of itself.[2] There is much more to reading a story, in other words, than attending to the grammatical rules required for its telling. Reading a story includes becoming aware of emergent new meanings that adhere to grammatical requirements but that mere expertise in grammar cannot specify.[3] Analogously, one may look at the cosmic story from many generalized scientific points of view, but scientific analysis alone cannot appreciate the drama going on inside.

One way to read the universe scientifically, for example, is as a series of distinct kinds of energy-flow, as Eric Chaisson's version of Big History does.[4] This is an illuminating "outside" perspective, but the universe is more than a set of recurrent physical regularities. From anticipation's dramatic point of view, when mind comes onto the scene, the whole universe becomes unpredictably new, and it does so without interrupting the fixed rules of thermodynamics. My point, once again, is that physical science, strictly speaking, cannot even see, much less understand, the shifts of dramatic meaning taking place inside the cosmic story. Physics, chemistry, and genetics can no more read what is really going on in the emergence of life and mind than your knowledge of the rules of grammar can specify the meaning of a novel you may be reading. The modern claim that the universe carries no dramatic meaning, on this analogy, is a product of archaeonomic "grammaticalism." It is the result of a fixation on recurrent physical routines and invariant rules at the expense of interest in emerging new narrative content. What this narrative content may be in the case of the cosmos, however, is a topic for succeeding chapters.

The third and most dramatic breakthrough in the self-unification of the universe has occurred in the emergence of God-consciousness. The axial religious sense of a unifying infinite transcendence is a development of cosmic importance, but we do not expect its arrival to have suspended or abolished the molecular, metabolic, and electrochemical patterns operative in the cellular life and central nervous systems of religious devotees. Expertise in physics, biology, or neuroscience is not enough either to record or rule out any dramatic cosmic significance in a universe awaking to a transcendent principle of unity. When we read the universe with the eyes of anticipation, however, the emergence of God-consciousness seems to give the whole universe an unprecedented dramatic new twist. When human minds begin to open themselves gratefully to an infinite horizon of rightness and to a transcendent principle of unity, the whole universe has become dramatically new for a third time.

In the birth of God-consciousness the cosmos consciously hands over its past and present plurality to a transcendent principle of unification. In religious persons and their communities the still dispersed cosmos acknowledges that the unity it seeks comes as a gift from beyond and not as a necessity arising from its own physical past. Archaeonomic grammaticalists, true to form, fail to notice. Unable to appreciate the dramatic cosmic changeovers that had already occurred in the emergence of life and mind, they are hardly prepared to grasp the cosmic significance of religion. They are conditioned to interpret religion as just one more way in which lifeless and mindless physical elements are falling together into biological and neurological patterns.

From the perspective of an inside cosmic narration, however, the axial breaking out of God-consciousness is no less dramatic than the earlier arrival of life and mind. The religious awakening to a unifying font of rightness lying beyond immediate grasp takes place without nature's having to alter the routine manner in which energy exchanges, chemical activity, organic processes, genetic rules of inheritance, or human brains function. The awakening of our universe to a transcendent principle of unity has overstepped no physical, chemical, biological, or neurological habits of nature. Measurements of brain activity may observe that religiously influenced meditation makes parts of the brain more noticeably active than they are during other kinds of experience. Yet these observations, whose significance has been in my opinion massively overblown, are trivial in the sense that they too remain blind to the dramatic cosmic significance of the arrival of God-consciousness. Even though scientific specialists in neuroscience, psychology, and biology may legitimately "explain" religion from within their accustomed disciplinary boundaries, the grammaticalist fixation of pure archaeonomy fails to notice the dramatic and unifying significance of the recent eruption of God-consciousness in cosmic history.

The Drama of Striving

Before the arrival of mind and religion, the emergence of life was the most significant development in the drama of the universe. With the arrival of life, as I noted earlier, came a whole new way of being, namely that of striving. In the emergence of organisms that endeavor to achieve certain goals, the universe became abruptly new. Life's capacity to strive ushered into the cosmos the note of drama, since living organisms may either succeed or (tragically) fail in the effort to realize their aims. The dramatic quality of life remains unnoticed, however, unless we look at living beings in terms of their capacity to succeed or fail. Looking at them only in terms of their physical and chemical makeup is not enough to let us in on the drama of life. We personally grasp the reality of living beings and distinguish them from inanimate processes only inasmuch as we, as striving beings ourselves, sympathetically identify with their capacity to make an effort, and hence at times to fail, in their endeavoring to achieve their goals.[5]

Subjective striving is something we humans undeniably experience immediately and palpably in our own lives. You are doing so right now, for example, in striving to reach a right understanding of this chapter. By virtue of any organism's capacity to strive, achieve, and possibly fail, the whole universe drapes itself in the curtain of drama. Whether it is an amoeba looking for nourishment, a gazelle trying to escape a predator, or a college sophomore devouring books in search of meaning and truth, the fact of striving is what gives life both uniqueness and unity. And it is the capacity for striving that gives the universe its dramatic quality.

Now that science allows us to read the universe historically, we realize that the entire narrative of nature, including its preliving epochs, is complicit in the spontaneous emergence of life—and that means striving—even if the initial occurrence of life is a single moment late in time and highly local in space. During the past century, the natural sciences have demonstrated with unprecedented precision how intricately the spatial and temporal transformations that occurred during the preliving phases of cosmic process are connected to the eventual bursting out of life, and hence of striving, in our terrestrial habitat. Cosmic and sidereal events that used to seem remote and irrelevant to what is going on here on Earth are now acknowledged to have been a necessary prelude to, and hence part of, the drama of life.

Life means striving, and striving introduces into the universe the dramatic notes of possible success or failure. But the existence of life itself depends on physical conditions and constants that had to be in place at the time of cosmic origins. Large-scale, intermediate, and subatomic facets of the universe all have mathematical values that are specific to the manufacturing of living cells and hence to the weaving of life and subjectivity into the tissue of a temporally irreversible cosmic performance. The long passage of time, spatial expansion, and burgeoning complexity of the natural world—as exposed by astronomy, geology, chemistry, biology, and other sciences—is integral to the drama of striving that we know as life.

The anticipatory religious longing for a saving transcendent rightness is perhaps the most adventurous instance of striving in the whole story of life. The axial arrival of God-consciousness is life's universal striving now concentrated in the unprecedented intensity of human longing to transcend the plurality from which the world has sprung. Through religious striving for communion with a transcendent unity, billions of years of cosmic activity, along with all the expansive enormity of space and emerging physical complexity, receive a new narrative coherence.

Skepticism

To those who are skeptical of such a proposal, I can only reply, first, that they may be reading the universe in an exclusively archaeonomic/grammaticalist manner and, second, that any purely physicalist reading of nature is inherently blind to the dramatic depths of the universe. Consequently, it is inevitable that, in a materialist reading, sooner or later questions about the veracity of religion, and especially God-consciousness, would arise. These questions have become increasingly heated ever since the birth of modern science four centuries ago. A typical complaint, for example, is that there is no "evidence" for the beyondness to which the idea of God makes reference. Isn't divine transcendence, in other words, a complete void? What kind of existence does it have, if any at all? Is not the whole idea of transcendent rightness a human invention born of childish wishing to escape the annoying constraints of purely physical existence?

Our three main ways of reading the universe provide sharply different responses to these questions. Archaeonomic naturalism, to start with, gives no ontological status to transcendence, and it firmly rejects the religious prospect of a decisive deliverance of either ourselves or the cosmos from final disintegration. Transcendence, in this perspective, is equivalent to nothingness. By contrast, the analogical vision, embracing an otherworldly optimism, thinks of the beyond as a timeless unity infinitely more real than the world available to everyday experience and scientific comprehension. Analogically understood, the invisible transcendent world is already a fully actualized and inherently unchanging unity. Finally, the anticipatory vision, with its vivid awareness of a still-emerging universe, interprets transcendence as the infinitely generous font of yet unrealized possibilities dawning on the horizon of the cosmic future. Transcendence, in this reading, is experienced only by anticipation. As such, it lies out of direct sight and seems unreal except to those who risk adopting a posture of expectation. To encounter it even fleetingly, anticipation instructs us, we must have acquired, by the grace of its dawning light, the epistemologically essential disposition of attentive, patient waiting —a stance that the Abrahamic religious traditions refer to as hope.

But is there a transcendent unity that unobtrusively invites the universe to become more? Is there in fact an indestructible rightness gathering the cosmos into a unity that preserves rather than dissolves all diversity? Is there an imperishable rightness that brings unity to all things, as mystics of many traditions have claimed? Or, in the age of science, should we not regard transcendence as an obsolete illusion? Granted, the sense of transcendence seems consoling and real to those who feel imprisoned by their present experience of the world, but is it reasonable to put our trust in it, as religion asks us to do?

Archaeonomy's answer, of course, is "no." From everything the pure naturalist can make out, the cosmos is destined fully by its mindless physical past to end in absolute death. Disintegration into the nothingness of sheer multiplicity is the inevitable final state of all being. Such a thought is initially sobering, but it is not as disturbing to cosmic pessimists as it may seem to religious believers. To some contemporary archaeonomic naturalists the now fashionable idea among cosmologists that the universe has no need of a transcendent origin and destiny, and that it "just happened" and "has no point," seems refreshing. Some scientific skeptics even feel a sense of liberation at the thought that no transcendent dimension of mystery encompasses or awaits the universe. Occasionally archaeonomic naturalists admit that, like all humans, they are genetically wired to be religious, but they consider the new scientific portraits of the universe big enough to satisfy their spiritual longing for expansive horizons.

Transcendence, in the archaeonomic reading, is a notion we humans have made up, and it is only right that we now admit it! The world open to scientific inquiry is interesting enough to fill our lives, they claim, without our having to put up with the entanglements of religion and theology.[6] The universe exposed by contemporary astrophysics does not need to have an origin and goal beyond itself. The universe of science is big enough to give humans plenty of room to live and breathe. Archaeonomic naturalists believe that the natural world is also rich and resourceful enough to satisfy their longing for meaning, goodness, truth, and beauty. Furthermore, the new cosmic story, having magnified their world-pictures far beyond anything available in the literature and myths of religion, is an improvement on the countless creation stories concocted by the prescientific religious imagination down through the ages.

Archaeonomic naturalists therefore insist that our human astonishment at the immensity and complexity of nature is misplaced. Wonder may be psychologically healthy, but religious wonder is a childishly inappropriate reaction to a universe that, at bottom, consists of a limited set of simple physical "laws" governing a multitude of physical elements and events. Religious talk about an incomprehensible mystery beyond the universe is a symptom of ignorance. The proper aim of science is to expel mystery for good, thus rendering religious wonder an obsolete sensibility. "Now that astrophysicists understand the physics of the sun and the stars and the source of their power," the late physicist Heinz Pagels declared, "they are no longer the mysteries they once were. People once worshipped the sun, awed by its power and beauty. In our culture we no longer worship the sun and see it as a divine presence as our ancestors did." Even though scientifically illiterate people "still involve their deepest feelings with the universe as a whole and regard its origin as mysterious . . . the existence of the universe will hold no more mystery for those who choose to understand it than the existence of the sun." Accordingly, "as knowledge of our universe matures, that ancient awestruck feeling of wonder at its size and duration seems inappropriate, a sensibility left over from an earlier age."[7]

In the final analysis, archaeonomic naturalists claim, the natural world shows no signs of being encompassed by or grounded in transcendence. To Albert Einstein the fact that the universe is comprehensible may be an incomprehensible mystery, but this mystery does not consist of a supernatural realm of being that lies beyond the physical universe available to ordinary experience and science. For contemporary scientific naturalists the physically accessible universe is all there is, and its inevitable destiny is to fall back eventually into separate atomic and subatomic bits. The sacred mystery of "God" that many religious believers claim to experience is a groundless human fabrication.[8] As the physicist Sean Carroll puts it, "If and when cosmologists develop a successful scientific understanding of the origin of the universe, we will be left with a picture in which there is no place for God to act." Truth-loving people, therefore, must suppress their inherited religious instincts and acknowledge the sheer givenness, simplicity, and atomicity of nature. Science, in this perspective, has made the idea of a transcendent personal deity altogether unnecessary.[9]

After officially claiming that the world available to science and ordinary experience is all there is, however, the archaeonomic reading of the universe forks off into two distinct branches. One offshoot is dark and sober, the other light and sunny. The sober archaeonomic reading, subservient to the metaphor that nature is driven by inviolable physical "laws," agrees that science has shown nature to be unremarkable, multiple, and monotonous underneath all of its improbable syntheses. Unable to discern any real drama going on inside the universe, archaeonomic naturalists insist that there is no exit from the impersonality of it all. Sober archaeonomic naturalists therefore repudiate any suggestion that the universe is a drama of awakening to rightness. And they are especially hostile to the idea of a self-revealing and saving personal God.[10]

This darker variety of archaeonomic pessimism, I should note in passing, leaves room for the kind of consolation that comes from huddling together with one's fellow humans to find a small degree of warmth in an otherwise cold and heartless universe. On rare occasions, moreover, a strict cosmic pessimist may profess to be experiencing a kind of solitary satisfaction in heroically facing up to the pointlessness of the universe. The number of scientific materialists who reach this exceptional mental state, however, is limited. Sober archaeonomic naturalism is simply not a space within which most human beings can live. It functions more as a theoretical limit than as the defining feature of a sociologically identifiable body of science-minded people.

Sunny archaeonomic naturalists, for their part, agree intellectually with their sober counterparts that the material universe is all there is and that there is nothing beyond what science can give us. They are more hospitable, however, to people who look for signals of transcendence and the prospect of surviving death. Life is often miserable, they admit, so we should not blame religious people, especially the scientifically illiterate masses, for looking for more-being beyond the natural world. Nevertheless, to the sunny naturalist, God-consciousness is incompatible with right understanding and needs to be outgrown. In my own experience, most materialist scientists and philosophers are inclined temperamentally to be sunny rather than sober.[11] Only rarely does one find the darker version of naturalism being rigorously thought through, much less lived out, with logical consistency.

Both the sunny and sober forms of archaeonomic naturalism reject the analogical vision of nature. They deny the existence of a perfect, completely changeless sphere of transcendence beyond the physical universe. Religious symbols, they assume, are cognitively empty. If symbols reveal anything at all it is not a transcendent rightness but the childish psyches of people who lack the courage to face a godless universe. Usually when archaeonomic naturalists talk about religion, they have in mind some version or other of the analogical vision. They consider otherworldly optimism unacceptable since it seems to diminish the significance of our lives within the natural world. For them it is not right to flee from nature into an imaginary supernatural world. Convinced that we humans have a purely this-worldly future, bleak though it is in the final analysis, archaeonomic naturalists think we should work together to make life on Earth better without calling on help from beyond and without hoping for any final cosmic fulfillment.

The analogical worldview, by contrast, having come into prominence during the prescientific period of human history, is appealing precisely because it holds out the prospect of human persons being rescued by a transcendent dimension of being from imprisonment in a decaying material world. The natural world exists as a temporary setting in which to work out one's personal salvation. Understood in this way the physical universe does not call for human efforts to transform it as though it had any lasting importance. Analogy, nonetheless, is not altogether insensitive to the value of nature since it reads the universe sacramentally. The lovely things in nature, in other words, may function as symbols pointing to a sacred "beyond." Sacramentalism puts us in touch with the transcendent goodness and beauty of the infinite. Nature's value here and now consists in part, then, of its being a window onto an incorruptible rightness that infinitely transcends the natural world. The physical universe, however, is in principle disposable in the long run.

Anticipation, on the other hand, seeks to plant a fresh religious sense of transcendence, including a new kind of God-consciousness, on the terrain of a universe still being born. To anticipation the transient beauties of nature are not reminders of an otherworldly realm of being that is already complete. Rather, they are intimations of a fullness of unity and beauty yet to be realized. Anticipation views the universe as a process of gathering the still scattered elements of the past and present into unprecedented syntheses not yet actualized. The idea of transcendence refers to the ultimate center, attractor, and goal of the emerging story we are calling the universe. Contrary to the analogical vision, anticipation proposes that what gets accomplished in world history is not to be discarded and abandoned in the end. Instead it is to be woven narratively and everlastingly into transcendent rightness itself. What goes on in the universe really matters, not only now but also forever.

Waiting

In their respective attempts to unify and understand the physical universe neither analogy nor archaeonomy is willing to wait. To anticipation, however, waiting for the universe to become an actual unity is essential not only to right understanding but also to authentic human existence and, as we shall see in Chapter 11, to human happiness. Waiting, however, is not just idly whiling the time away or putting up passively with long periods of time. Waiting is attentive and active. It is inseparable from the act of awakening. In an unfinished universe, vigilant waiting is the most realistic posture that seekers of right understanding can possibly assume. In the Abrahamic traditions, the anticipation of infinite rightness and transcendent unity takes the form of waiting in hope for a new and surprising future. Hope that is seen, however, is not hope, as the Apostle Paul instructs an early Christian community (Romans 8:24). The fulfillment of hope is something for which we have to wait.

Patient waiting, however, is essential to religion worldwide, and not just to Judaism, Christianity, and Islam. Religion everywhere is a matter of expectation more than possession. Think, for example, of how the anticipation of final liberation from the cycle of rebirths subtly informs Buddhist expectation, leading to a vast literature and many religious practices concerned with cultivating the right disposition for approaching *nirvana*. Essential to all religion is the longing for a decisive deliverance from wrongness, but such deliverance does not happen by magic or decree. Religion, especially in the axial traditions, means having to wait.

Mahayana Buddhism, though not concerned with God-consciousness, offers a riveting example of religious waiting. This popular branch of Chinese and Japanese religion idealizes the figure of the *bodhisattva*, one who having reached the threshold of nirvana forgoes final deliverance and is willing to wait until all of life is ready to join in. A famous Mahayana text pictures the bodhisattva as having these sentiments:

> All creatures are in pain. All suffer from bad and hindering karma. All that mass of pain and evil I take in my own body. Assuredly I must bear the burden of all beings for I have resolved to save them all. I must set them all free, I must save the whole world from the forest of birth, old age, disease, and rebirth. . . . For all beings are caught in the net of craving, encompassed by ignorance, held by the desire for existence; they are doomed to destruction, shut in a cage of pain. . . .

It is better that I alone suffer than that all beings sink to the world of misfortune. There I shall give myself into bondage, to redeem all the world from the forest of purgatory, from rebirth as beasts, from the realm of death. I shall bear all grief and pain in my own body for the good of all things living. I must so bring to fruition the root of goodness that all beings find the utmost joy, unheard of joy, the joy of omniscience.[12]

The religious encounter with transcendence cannot occur apart from patience. Religion, in the anticipatory vision, means allowing time for things to grow and ripen. Pierre Teilhard de Chardin illustrates what I mean when he gives the following spiritual advice to his fellow believers, most of whom have grown up, as he did, with analogical religious sensibilities:

Above all, trust in the slow work of God. We are quite naturally impatient in everything to reach the end without delay. We should like to skip the intermediate stages. We are impatient of being on the way to something unknown, something new. And yet it is the law of all progress that it is made by passing through some stages of instability—and that it may take a very long time. And so I think it is with you; your ideas mature gradually—let them grow, let them shape themselves, without undue haste. Don't try to force them on, as though you could be today what time (that is to say, grace and circumstances acting on your own good will) will make of you tomorrow. Only God could say what this new spirit gradually forming within you will be. . . . Accept the anxiety of feeling yourself in suspense and incomplete.[13]

Waiting, in short, is fundamental to an anticipatory spirituality. Teilhard is exceptionally aware that religion, like the unfinished cosmos that carries it, is always incomplete. Religion seeks perfection, but it is never perfected. It is always on the way, and it can go astray. Like the universe, religion is an ongoing project. Religion has a dark side, but it also has a future, so it cannot be fully or fairly defined, any more than anything else, in terms of its past performance or present deficiencies alone. Nor can its current grasp of transcendent rightness be expected to be anything more than provisional.

Thus the fact of an unfinished universe liberates religion from the expectation that, to be worthy of our attention, it must already be polished to perfection. Acknowledging that religion, as part of an unfinished universe, still has a future, therefore, could be a fruitful starting point for interreligious conversation today. The unity of religion lies not in exclusivism, nor in the recovery of a primordial tradition, nor in reducing it archaeonomically to physical or biological factors. Rather, the quest for religious unity in an unfinished universe begins with the shared anticipation of a transcendent rightness that is both real and waiting to be realized.

Transcendence, in summary, is already presupposed by the striving we call life. From a human point of view there must be an inexhaustible horizon toward which our own striving has to move if we are to experience the fullness of life. A sense of transcendence, therefore, is not a luxury but a basic requirement of vitality, whether acknowledged or not. Without striving, there is no life, but without an unrestricted field of new possibilities there can be no room for limitless striving, and hence no room for real aliveness. Unlike nonhuman forms of life, human vitality requires a conscious openness to the ever-moving horizon of endless transcendence. To anticipation this transcendence lies not so much up above as up ahead of an emerging universe. Since the still-dawning horizon of infinite rightness lies mostly beyond our grasp, however, we can currently refer to it only by means of a rich and privileged kind of discourse, namely, that of symbols. Ever since the earliest beginnings of humanity's terrestrial sojourn it has been the function of religious symbolism to open consciousness up to the unrestricted horizon that human vitality requires. Apart from enlivening ways of symbolizing transcendence we wither and die. In the following chapter I explore the meaning of religious symbolism in the setting of our new awareness that the cosmos is still coming to birth.

CHAPTER SEVEN

Symbolism

We could easily be made to believe that nothing had happened, and yet we have been changed, as a house is changed into which a guest has entered. We cannot say who has come, perhaps we shall never know, but there are many indications to suggest that the future is entering into us in this manner in order to transform itself within us long before it happens.

—RAINER MARIA RILKE

INDESTRUCTIBILITY AND TRANSCENDENCE — qualities that religion commonly associates with rightness—share the trait of being hidden from both science and ordinary experience. What religion takes to be eminently right, real, good, true, and beautiful remains out of sight and out of mind, or at least mostly so. Devotees of archaeonomy, of course, think this invisibility means that nothing is there to begin with. To analogy, the mark of hiddenness is a sign not of the unreality but of the inherent incomprehensibility of rightness. Absence of evidence is not evidence of absence. To anticipation, however, the obscurity of rightness has to do with the fact that the universe is not yet fully awake. Rightness is hidden because, even though it is present and real, there is also something about it that has yet to be realized.

It is a mark of impatience, therefore, when archaeonomic naturalists demand that rightness, in order to win our recognition, needs to make itself fully manifest to human comprehension here and now.

Religion, it appears, has to protect rightness from being cheapened by easy accessibility, so our references to it must be indirect rather than direct. Hence religion clothes rightness in the irreplaceable apparel of symbolic expression. A symbol, generally speaking, is a way of representing something indirectly by pointing to something else directly. A symbol, however, is not the same thing as a sign. A sign, such as a stop sign on a street, is an object to which a contrived meaning—for example, that the color red signifies the need to stop—is assigned arbitrarily. Connecting "stopping" to redness is a sociocultural convenience, not a natural requirement. A symbol, on the other hand, is not arbitrary. A symbol naturally participates in or, as I shall propose below, anticipates that to which it points.[1]

"Symbolism" is a term I use here, for convenience's sake, to include all the indirect modes of reference that religion employs, including metaphors, myths, rituals, and analogies. Symbolism is the indispensable language of religion. Religion by necessity refers to transcendence indirectly by pointing directly to special qualities found in physically available things and persons in the perishable world. So the psalmist speaks of God as a rock, and of divine protection as a fortress (Psalm 71). Such images communicate to devotees a feeling of confidence that, in spite of suffering, death, and other threats, their lives remain connected to something everlastingly trustworthy.

Finite objects that seem good and beautiful, as well as persons whose lives are exemplary, can function symbolically to give religious devotees a sense of being encompassed by an infinite goodness, beauty, and fidelity. In Christianity, for example, the person of Jesus who is called the Christ functions as the main symbol of transcendent rightness. Christians are instructed not to think about God at all without thinking simultaneously about Jesus as presented in the New Testament writings and Christian tradition. So also in Indian religion, ultimate reality and meaning are symbolized through countless colorful images, persons, and objects that point beyond themselves to the transcendent rightness of Brahman.

Almost anything can symbolize—and thus become more or less

transparent to—the elusive reality of rightness that lies beyond direct experience. The ritual of being washed externally by water, for example, can symbolize the experience of being refreshed inwardly by an encounter with transcendence. This accounts for why ablutions abound in religion. The transient experience of illumination that occurs during the rising of the sun each day can represent symbolically the dawning of an inextinguishable rightness in one's life, and this is why the symbol of light is nearly universal in religious traditions. Additionally, the everyday experience of awakening can point symbolically to the hidden religious drama of interior transformation. Awakening to the light of day, therefore, is also a nearly universal symbol for being challenged by the encounter with the mystery of rightness.

Religious traditions bring devotees to symbolic awareness of ultimate reality in culturally distinct ways. Popular religion clothes its intuitions of rightness in locally familiar personal imagery such as mother, father, friend, spouse, shepherd, helper, warrior, or liberator. Religion also gives rightness proper names such as Christ, Krishna, and Gautama. In most religious traditions, in fact, personal symbols are essential. If rightness were less than personal, people of faith assume, it could not move or attract us at the center of our being. Impersonal objects may be intriguing, but they do not flood a person's life with meaning in the face of disaster and death. For most religious believers, therefore, the source and goal of their lives has to be a "Thou" and not merely an "It."[2] Since "Thou-ness" entails subjectivity, and subjectivity is always hidden inside, religion's attributing personality to ultimate reality is an important way of protecting rightness from being lost in the impersonal world of mere objects. Accordingly, whenever subjectivity seems unreal, as it does to contemporary archaeonomic naturalists, so also do notions of a personal deity. The expulsion of subjectivity from nature has been a necessary intellectual prelude to the debunking of every idea of divine personality.

Symbolism and the Meaning of Faith

The incomprehensibility of rightness, as I have just noted, demands that the primary language of religion be indirect rather than direct.

Human minds and hearts cannot grasp rightness directly, but they may have an awareness of being grasped by it. A good name for this awareness is "faith." "Faith" (in Greek *pistis*) is a biblical term meaning steadfast, patient confidence in God, but here I want to associate it more widely with the trust that comes from all religious experiences of being grasped by rightness.[3] In this generalized sense faith everywhere comes to expression first through symbols, rituals, myths, stories, analogies, and metaphors. Only later—as in the present book—does it become the topic of systematic reflection. Religious traditions agree that being grasped by rightness in a conscious way is not inevitable or automatic. It requires rites of passage, along with the practice of meditation, prayer, and good works. Neither normal awareness nor science can grasp the content of faith. Through symbolic acts and expressions, however, the transcendent mystery of indestructible rightness can awaken and transform those who are willing to risk a life of faith.

For most people, faith requires personal symbols of transcendence. Nevertheless, faith is not restricted simply to searching for intimacy with a hidden subject or "Thou." Most religious traditions, including Hinduism, Daoism, Buddhism, Judaism, Christianity, and Islam, are aware that rightness is not fully captured by personal imagery. Consequently, they also use nonpersonal or superpersonal language when referring to absolute reality. The term "rightness" as used in this volume is an example of what I mean. Not everything about absolute, transcendent reality can be captured in first- or second-person discourse. Abstract or neuter language is also necessary to emphasize the beyondness, omnipresence, and indestructibility of rightness. So in the wide world of religion rightness has been called by such nonpersonal names as *dao, ma'at, wisdom, logos, dharmakaya, being-itself, infinity, ground, depth, truth, goodness,* and *beauty.*

When religion refers to rightness, however, whether nonpersonally or personally, it does not expect to grasp the content of faith in the same experimental, objectifying way that science deals with things it investigates. In religious traditions the attempt to master and control rightness—trying to bottle and cap it, as it were—is considered wrong. Muslims call such an attempt at mastery *shirk,* Jews and Christians refer to it as *idolatry,* and Buddhists name it *tanha.* To approach the content of religion in an exclusively scientific or archae-onomic way seems blasphemous to people of faith. Expecting right-ness to be captured by ordinary experience, or nowadays by the tools of science, would amount to an abasement of faith.

Consequently, especially since the axial period, religion protects faith and its content by emphasizing the necessity of silence. The infinite transcendence to which religious symbols point is mostly unspeakable, and it is the function of prayerful silence to remind us of the inadequacy of all references to it. Terms such as God, *dao*, and *nirvana* refer to a rightness that is so far out of our grasp that genu-ine piety at times must give way to a contemplative stillness that protects the goal of religious longing from undue diminishment by ordinary language and symbolic reference. Accordingly, to insist on a precise or exact meaning of religious symbols is always inappropri-ate. Religion does not have to obey the impatient requirement, es-pecially dear to modern scientism, to make everything subject to clarification by physical analysis. To insist that the symbolic content of faith be translated into mundane, scientific, or philosophical lan-guage would be to miss its meaning entirely. As the Dao de Jing says:

> Gaze at it; there is nothing to see.
> It is called the formless.
> Heed it; there is nothing to hear.
> It is called the soundless.
> Grasp it; there is nothing to hold on to.
> It is called the immaterial.
> Invisible, it cannot be called by any name.
> It returns again to nothingness.[4]

Rightness is dawning always, but it hides from easy publicity. Yet even in its hiddenness and humility it is all-powerful. Rightness influences the world as a consequence of, and not in spite of, its nonavailability:

> Thirty spokes are joined at the hub.
> From their non-being arises the function of the wheel.
> Lumps of clay are shaped into a vessel.

From their non-being arises the functions of the vessel.
Doors and windows are constructed together to make a
    chamber.
From their non-being arises the functions of the chamber.
Therefore, as individual beings, these things are useful
    materials.
Constructed together in their non-being, they give rise to
    function.[5]

Wu Cheng (1249–1333) interprets this text: "If it were not for the empty space of the hub to turn round the wheel, there would be no movement of the cart on the ground. If it were not for the hollow space of the vessel to contain things, there would be no space for storage. If it were not for the vacuity of the room between the windows and doors for lights coming in and going out, there would be no place to live."[6] We can appreciate the humble effectiveness of rightness (*wu wei*), then, only if we first allow our own lives to be transformed—that is, made receptive, humble, and grateful—by the unobtrusive influence of the dao. Archaeonomic naturalism, by contrast, is fixated on the spokes, clay, walls, and door frames—that is, on what is easily visible and controllable. It has no inkling, therefore, of the humble and hidden rightness that gives being and meaning to everything but that remains humbly in the background. It has no room for symbolic reference to a dimension of transcendent importance whose reality can be acknowledged only by our first allowing it to take hold of us.[7]

Religious Symbolism in an Unfinished Universe

Religion, then, makes reference to transcendent rightness chiefly through a combination of silence and symbolism. It refuses adamantly to embrace the modern ideal of absolute clarity and intellectual control over its subject matter. Religious symbols can never be translated fully into the direct kind of reference that modern scientism and rationalism demand. If the content of religious symbols could be fully conveyed in ordinary language or scientific discourse, there would be no awareness of participating in a transcendent rightness at all. Indeed, there would be no religion.

What, though, does religious symbolism look like in the setting of a universe that is still coming into being? To answer this question, once again we need to look even more closely and comparatively at our three ways of reading the universe and religion. Let us begin by providing only a sketch of how each would interpret religious symbols, and then follow this up with a more detailed treatment.

Briefly, the archaeonomic reading interprets religious symbols as purely imaginative fabrications by which humans try to escape from or adapt to a pervasively pointless universe. To archaeonomy the inherently meaningless cosmic process is all that really exists, so any attempt to discover symbolic meaning in it is a cover-up rather than an uncovering. The way of analogy, ignoring the fact that the cosmos is still unfolding in new and unpredictable ways, understands religious symbols as "sacramental" reminders of a perfect, change- less, and sacred world beyond time and space. Anticipation, however, interprets religious symbols as ambiguous advance announcements of a rightness that is still dawning. Religious symbols are the revela- tion beforehand of what has not yet fully come to pass.

Religious symbols, in this third perspective, are neither illusory projections nor imperfect analogical reminders of an eternal perfec- tion outside of time. Rather, they are installments of a rightness that is both real and waiting to be realized. The cognitive slipperiness of symbols is not a good reason for suspecting them (archaeonomically) of being mere illusions. Nor is it an excuse for interpreting them (analogically) as the hide-and-seek playfulness of a perfection already realized. Rather, from an anticipatory perspective, the ambiguity of religious symbols is a function of their being part of an awakening universe. Religion is symbolic, in other words, because the universe is still coming into being. A realistic appreciation of religious symbols therefore requires that they be approached with the same patience that allows the universe time to become coherent. Let us now examine more closely each of the three ways of reading religious symbolism.

Archaeonomy. The religious use of symbolic expression is the source today of considerable ridicule by scientific naturalists. To contemporary critics of religion such as the New Atheists, faith is nothing more than "belief without evidence." For them the univer- sal religious claim that what is most real can be approached "only" metaphorically, or by way of symbols, is indicative of what is most wrong with religion. To the pure naturalist, religion's symbolic mode of reference is a mask for human ignorance and an excuse for self-deception. Religion, to be believable, should present visible "evidence" that rightness is real—or, in a theistic context, that God exists. Religious symbols, according to archaeonomic assumptions, are pure fiction, and sadly, the faithful everywhere fail to acknowledge their emptiness.[8]

To the evolutionary naturalist, religious symbols are at best adaptive illusions invented by humans, or more remotely by populations of human genes, and projected onto an inherently blank cosmic canvas. There is nothing or no one "out there" to which religious symbols for rightness could possibly be pointing. Nature is an immensely long, exclusively "outside" affair. Beneath its filmy narrative surface lies a pointless process of shuffling and reshuffling atoms, molecules, and genes. Religious symbols, in this interpretation, are simply the main way in which scientifically ignorant people have blinded themselves to the pointlessness of it all. Humans, Michael Shermer writes, are addicted to looking for patterns where there are none, and religious symbols are all projections of purely imaginative schemes of meaning onto a material world that, underneath it all, remains indifferent to our personal longings.[9] Religious symbols in this materialist interpretation are products of human brains sculpted over millions of years by natural selection, a process that is reducible in turn to a deterministic arranging and rearranging of elemental physical units across the enormity of space and time.

All the materialists we have met so far—Simon Blackburn, E. O. Wilson, Richard Dawkins, Daniel Dennett, Alex Rosenberg, David Papineau, and others—are sure they are right in claiming that religious symbols are referentially empty. It is not hard to see, however, that in every claim they make—for example, that science alone yields right understanding—they are in fact measuring their beliefs tacitly against a background of rightness that they also have to refer to symbolically. A hidden dimension of idealized rightness keeps seeping into the archaeonomic naturalists' universe through wide-open cracks in their own cognitive performances. They refer to rightness unconsciously by way of their own cherished system of symbols. Although they explicitly deny it, an elusive transcendent rightness looms quietly on the horizon of their own inquiries and claims. They tacitly appeal to it in every judgment they make about the truth of their own claims and the falsity of religious symbols and myths.

Since rightness even for the materialist cannot be grasped directly, it too must be referred to by way of an unacknowledged set of symbolic and mythic expressions. For the pure archaeonomic naturalist the term "matter" itself functions symbolically as a name for true being. Mindless "matter" stands for what is ultimately real. "Matter" is a symbolically imprecise name for the ground and destiny of all things. "Matter," a term that has never been clearly defined, points symbolically to the indestructible source and final repository of life and mind. Materialism then is a mythic-symbolic worldview that proposes to answer the perennially big human questions no less comprehensively than other belief systems. Materialism tells its devotees where everything comes from, where everything is going, what is real, and what is not. Like all myths of origin and destiny, materialism draws strict boundary lines around the world to keep chaos at bay. It becomes especially defensive whenever it encounters the annoying cognitive raggedness of religious symbolism. It relieves its own anxiety in the face of uncertainty by reducing the world to what is mathematically comprehensible. In doing so, materialism generates its own set of ethical guidelines. Its first commandment: Do not accept as right any symbolic claims for which there is no physical evidence and potential scientific explanation!

Archaeonomic naturalism, therefore, turns out to be just one culturally limited mythic-symbolic vision of the world among others. Like all myths of origins, it seeks to specify (etiologically) how observable things came to be the way they are. Like other myths of origins, it tells highly symbolic stories—interwoven with physical and evolutionary content—about how things came to be. In this respect its accounts are not unlike other "archaistic" myths of origins that have been narrated by humans throughout our journey on Earth. When the novelist Ian McEwan's character Perowne insists that the new evolutionary story, unlike all others, has the "unprecedented" merit of being true, this claim is not at all unprecedented in its exclusivist trust that one narrative alone can put us in touch with absolute reality.[10]

True to their myth, all the materialists I have cited so far have uncritically swallowed and digested, in advance of their pronouncements about religion, the scientifically indemonstrable belief that only what is available to scientific inquiry can be called real and that every prescientific worldview is therefore "mythic" in the vulgar sense of being false. Finally, archaeonomic naturalism has its own gallery of heroic individuals who exemplify true courage by unashamedly demystifying the extravagant symbolic claims of religious traditions. Thus materialists are not so different after all from religious humans in anticipating a horizon of rightness and referring to it symbolically. Even in the act of denying transcendence, the light of rightness keeps dawning behind their backs, generously illuminating everything they seek to understand, yet remaining completely unnoticed.

As we have seen, the archaeonomic naturalists' first step in debunking religious symbolism is to ignore the reality of their own subjectivity, the place where rightness is registered. The strictest archaeonomic naturalists—sometimes referred to as "eliminative materialists"—even believe that subjectivity (and this would have to include their own) is an illusion.[11] They deny both the hiddenness and the transcendence of rightness by first refusing to acknowledge fully the hidden interiority operative in their every act of understanding and judging. Resting their entire understanding of the world on belief in the mindless fixity of the physical past, they trust—without evidence—that whatever takes place in nature and history will upon analysis turn out to be nothing more than the world's oldness decked out in the false attire of newness. The first article of their creed is that, hidden behind all transient appearances, the whole of being is at bottom nothing more than electrons, neutrons, quarks, and other subatomic entities.[12] Conventionally religious symbols, according to the archaeonomic way of reading things, are revelatory not of a dimension of rightness still dawning but of deluded human wishes projected onto the mindless physical elements that have been present from the beginning and that have now been exposed by scientific investigation. I should add here that archaeonomic assumptions about the emptiness of symbolic consciousness also underlie most contemporary versions of Big History, accounting for its inability to integrate interiority, including the story of religious awakening, into its narratives of the universe.

Analogy. To be sure, if nothing existed beyond the ambit of scientific objectification, there would be nothing to the religious sense of rightness. Religious symbols would indeed be purely human inventions and/or adaptive fictions. Transcendence, interiority, and everlastingness would be names for nothingness. Analogy, however, takes religious symbolism to be the most realistic of all human modes of reference. It locates rightness, conceived of as a primordial plenitude of being, in a changeless sphere of perfection that religious awareness here and now can contact, at least dimly, only through symbols known as "sacraments." Through symbols of the sacred, many of which come from our experience of nature, the essential hiddenness of eternity momentarily breaks into finite space and time where it becomes incarnate, albeit ambiguously, in the immediately visible world.

The analogical (sacramental) vision assumes that we can gain an imperfect but redemptive relationship to rightness through symbolic rituals such as immersion in water, anointing with oil, partaking of holy meals, or the sanctioning of mating by marriage ceremonies. Through a symbolic understanding of natural phenomena, religion connects us to what is most right and most real but outside the spatiotemporal world. People who have not undergone religious transformation—and today that would include most occupants of the science-savvy intellectual world—are unqualified to read the cosmos in its symbolic depth. Without going through an unsettling personal religious initiation, analogy instructs us, we shall all remain literalists, content with a surface reading that denies, ironically without scientific evidence, that anything real and significant could be hidden in principle from the sweep of scientific inquiry.

To the religiously initiated, special objects, persons, and events are symbolic windows that let in at least a little light from the hidden domain of sacred completeness. For centuries the visible star-studded heavens served, to people properly disposed, as a sacramental window opening onto an infinite beauty. To those whose interior lives were untransformed, however, even the heavens remained shrouded in darkness. In preliterate religion, sacraments derived from the wonders of nature gave those who were properly prepared an especially

vivid sense of a spiritual world extending beyond the mundane. Rightness was revealed through the sun, sky, earth, rivers, oceans, seasons,
animals, fertility, floods, and storms. Human sexuality, above all, was
a symbolically charged medium for religious connection to what is
most real. Nowadays, to be sure, sex has been secularized, but in
the history of religion, sexual experience and instances of fertility
have often functioned symbolically to put people in touch with a
hidden power of renewal that breaks through momentarily and then
retreats.[13]

Religion cannot exist or be passed on from one generation to
another without symbols, even today. Yet as religion has developed
over the past three thousand years, some of its leading visionaries
have come to suspect that a purely symbolic or sacramental reading
of nature can be dangerous. If taken too literally, symbols can shrink
rather than reveal the liberating power of rightness. Consequently,
to protect rightness from the reductionism known as "idolatry,"
there developed in both Eastern and Western religious thought an
unprecedented skepticism about the adequacy of symbols and sacraments to represent ultimate reality. Deliberate silence then increasingly became one of the marks of authentic religion.

Archaeonomic naturalism has its remote origins in the religious
distrust of sacramentalism that became a major aspect of axial religion centuries ago. Modern secularism in general derives its distinctive historical identity from carrying the religious need for silence
to a new extreme. Repudiating sacramentalism now in the name of
science rather than silence, contemporary secularism fosters an intellectual climate that replaces symbolism with literalism and what I
earlier called grammaticalism. Materialists claim that the real world
consists ultimately of physical elements and impersonal "laws." No
indestructible dimension of rightness exists to which religious symbols could possibly be pointing. Consequently, if our ideas cannot be
stated clearly and directly, in the way archaeonomy requires, we must
be covering something up. Symbolic consciousness is concealing the
real world beneath layers of human fabrication that we may now fear-
lessly remove. In its more sophisticated versions (for example, those
of Freud, Marx, and Nietzsche) secularism suspects that the whole
body of religious symbolism is not only fiction but also a mask for

life-denying and dehumanizing human desires.[14] It is only right, therefore, to let go of religious symbolism altogether.

Analogy, however, responds to the secularists that a properly religious practice of silence is not a denial of transcendence but a serious affirmation of it. The archaeonomic demystification of symbols, on the other hand, is not only an instance of blatant literalism but also a symptom of the arbitrary modern expulsion of subjectivity from the realm of true being.

Anticipation. Analogy rightly resists the reduction of reality to what can be mastered by the human mind, and especially by the modern obsession with clear and distinct ideas. From a Daoist, Hindu, or classical theological perspective, the modern urge to replace symbolism with clear and distinct ideas is a product of the will to power rather than of the humble desire to understand things rightly. Analogy correctly observes that if religious symbols were ever replaced by literalist clarity, the invisible rightness that nourishes intellectual, moral, aesthetic, and religious awareness would be lost to consciousness altogether.

It is now possible, however, to have new thoughts not only about nature and religion but also about religious symbolism if we situate the last inside an unfinished universe. The scientific discovery that the cosmos is still in the process of emerging allows for a revolutionary shift in human understanding of both symbolism and transcendence. The transcendent rightness awakening the cosmos, as noted in the preceding chapter, may be thought of not as a "timeless present" to be contacted only by withdrawing from nature but as an "anticipated fullness" to be approached by patient and attentive expectation. Rightness is both present fact and waiting to be realized. If rightness were exclusively a present fact, then what happens in time would not matter—since everything important will already have been accomplished from all eternity. The cosmos in that case would not be a meaningful story but instead a series of moments adding up to nothing more than what has already been. At the other extreme, if the ideal of rightness were completely made up by human subjects, then the cosmic process would be aimless.

Anticipation seeks to avoid both the analogical sense of time's futility and the archaeonomic impression of cosmic aimlessness. This avoidance is reflected once again in anticipation's novel understanding of religious symbolism. Symbols are neither fully revelatory of rightness nor altogether empty of significance. Religious symbols reflect the tension between the human experience of rightness as a present lure now awakening faith on the one hand, and the sense that the awakening is incomplete since rightness has not yet fully arrived on the other. Symbols are responses to an invitation to rightness but also openings to its fuller visitation. Anticipation appreciates religious symbols not only because they link us to transcendence presently but also because they are installments of a rightness still dawning. They are, as Rilke puts it, "indications to suggest that the future is entering into us . . . in order to transform itself within us long before it happens."[15]

To the archaeonomic naturalist, in summary, religious symbols are purely human creations devised to escape from an aimless universe, but to anticipation they are intimations of a consummation that is yet to be. To analogy, the present obscurity in religious symbolism may be a reflection of the world's temporal fall from an original revelatory transparency, but to anticipation the ambiguity of religious symbols is indicative of a still-awakening universe's present incompleteness.

Is Religion a Projection?

Since their light is obscured by long shadows, it has become increasingly fashionable in the age of science to treat religious symbols as escapist projections of human wishes onto the void of an inherently meaningless universe. The projection theory maintains that religious symbols are not revelatory of a hidden order of rightness, as religious people believe, but are illusions invented by weak mortals to hide from themselves the aimlessness of the universe. One of the most influential proponents of the projection theory, at least until recently, was the atheist Sigmund Freud. Freud rooted his projection theory in what he called psychoanalysis, a relentlessly archaeonomic method of investigating human interiority. Freud's theory still has some appeal, but archaeonomic naturalists and students of Big History these days prefer to go farther back in time, following the "outside" history of life as laid out by Charles Darwin, in search of the more remote evolutionary roots of symbolic consciousness. So-called evolutionary psychologists and evolutionary anthropologists now claim to be digging deeper than Freud into the biological and hence genetic origins of our symbolic propensities.[16]

The important point here is that in all of its many versions the projection theory will not work unless it takes for granted the fundamental aimlessness of the universe. Both psychological and biological versions of the projection theory understand the natural world as a blank screen onto which religious people project illusory symbolic fantasies. An anticipatory perspective, on the other hand, can give a realistic reading of religious symbolism that is also completely consistent with new scientific discoveries of an unfinished universe. Anticipation is fully aware of the ambiguity of symbols, but it protects them from being dismissed as pure projection. It reads religious symbolism as both the expression of human longing for narrative coherence and, at the same time, a self-unveiling of that coherence in advance. Even in all its ambiguity, religious symbolism *is* the universe now inviting us to consider seriously once again the possibility that it has a purpose.

CHAPTER EIGHT

Purpose

Adopt the pace of nature: her secret is patience.

—RALPH WALDO EMERSON

RELIGION GIVES PEOPLE CONFIDENCE that their lives are worth living and that the universe is here for a reason. To many thoughtful people, however, developments in the natural sciences have made the cosmos seem pointless. Cosmic pessimism, the belief that the universe has no goal or overall meaning, has been around since antiquity, but today it claims the support of physics, biology, and cosmology. Cosmic pessimists allow that living organisms including human beings have specific aims and act purposively, but this does not mean that the universe overall has a purpose. To scientifically educated people the widely shared religious sense that the universe is rooted in rightness and permeated by meaning usually seems now to be an unsupportable creation of the human imagination.

Nevertheless, since nature has undeniably given rise to conscious, meaning-seeking beings, it may not be unreasonable to ask whether, in producing such exceptional outcomes, the cosmos has to be deemed pointless, no matter how long the process has taken. In its billions of years of existence, we now know that the Big Bang universe—or the possible totality of worlds to which it belongs—has been undergoing almost continuous transformation. It has produced, among other splendid results, beings endowed with a restless longing for meaning and truth. Is it altogether certain that the universe that is now slowly awakening lacks any overarching purpose?

"Purpose," as I understand it here, means the realizing of rightness. In order for your own life to have purpose, for example, it would have to be lived in pursuit of something undeniably right. The goal of a purposeful life might be that of contributing to the well-being of nature, the establishment of peace among nations, watching out for the happiness of your neighbor, or perhaps increasing the store of human knowledge and the intensity of beauty. People involved in such pursuits generally experience their lives as purposeful. Is it possible, then, that something analogously purposeful is working itself out in the universe at large?

Cosmic pessimists firmly deny it. They are certain that the universe has no lasting importance or meaning. Usually they consider the question of cosmic purpose silly. It is enough, they say, that their own lives have meaning, and it makes no difference whether the whole scheme of things has a point. Nevertheless, as we have seen from the start, science is now clarifying, as never before, the intricate way in which each conscious subject is tied into the whole cosmic story. It is not as easy as it used to be to separate our interior existence, including our longing for personal meaning, from the natural world that gave birth to us. It could make a big difference in our quest for the meaning of our own lives if, upon examination, it turns out that the universe of which we are a part is not a pointless process. Accordingly, it may not be a waste of time to wonder whether our personal quests for meaning have the backing of the universe.

Analogical religion affirms our personal significance, but it attributes no lasting meaning to the cosmos. Anticipation trusts that something of imperishable significance is going on in the universe, but that we can encounter it now only by patient expectation. Archaeonomic naturalism, however, finds no permanent meaning either in our own personal lives or in the universe. It allows that human belief in purpose may be psychologically soothing and biologically adaptive, but it insists that there is no "evidence" that the universe is in fact awakening to everlasting meaning, truth, goodness, and beauty.

Even though they are pessimistic cosmically, however, this does not mean that archaeonomic naturalists are pessimistic psychologically. Most of them live normal, relatively happy lives. They generally carry on *as if* their lives have a meaning. Like other people, they raise families and work hard to ensure that they and their children will have satisfying careers. They love life and are eager to save planet Earth for future generations. They get involved in projects that enhance the quality of existence in their communities and workplaces. They assume that individuals can find meaning enough even if the cosmos provides no final support for their aspirations. Some even profess to be more than satisfied that the universe has no purpose since this allows them to fill in the blankness with their own meanings and values. "I, for one, am glad that the universe has no meaning," says the philosopher E. D. Klemke, "for thereby is man all the more glorious." A meaningless universe, he goes on, "leaves me free to forge my own meaning." Similarly, the late paleontologist Stephen Jay Gould declared that the pointlessness of the universe is not a reason for personal despair but an opportunity for us to realize our own special human significance. The cosmic meaninglessness exposed by Darwin's picture of life, for example, can now be colored over, he wrote, "with our own meanings."[1]

To archaeonomic naturalists, if purposiveness exists anywhere, it is not in the universe overall but only in a small slice of it known as the human central nervous system. Of course, the various body parts of living beings have a functional purpose. Wings and legs have the purpose of locomotion, and hearts of pumping blood. All living beings seem to be carrying out at least limited aims. We humans, however, are distinctive since we are said to have "purpose on the brain."[2] We are by nature meaning-seeking beings. Nevertheless, cosmic pessimists insist, a time will come in the far distant future when no traces will remain of life or of purposeful organs and meaning-seeking minds. While religious citizens of Earth look longingly toward a climactic, liberating communion with an indestructible rightness to ground their sense of purpose, cosmic pessimists insist that neither reason nor science provides support for such an expectation, and so it is not right to hope for any final redemption or fulfillment. Religious belief in immortality may give meaning to the lives of millions of people, they concur, but millions of people may be wrong.

The Indestructibility of Truth

Even in saying so, however, cosmic pessimists ironically fail to consider fully what it means that they are deeply devoted personally to the quest for right understanding—that is, to the search for truth. They seem to find a hidden meaning in the pursuit of right understanding even if finding the truth means having to conclude that the cosmos is absurd. The "outside" world available to scientific inquiry is certainly all there is, they claim, and it lacks any overall purpose, but facing this truth brings them an interior satisfaction that undeniably ennobles their lives. Unconsciously they allow themselves to be carried away by rightness, at least as they understand it. Even to cosmic pessimists there can be no absolute denial of indestructible rightness, only a formal limiting of it to what can be grasped by scientific method.

Science, as I am taking for granted throughout this book, undoubtedly illuminates our world, and I follow it faithfully as far as it goes, but its bright lights shine only on the outside, not on the inside. Science's beams, moreover, fall first on what has already happened and shed very little light specifically on future cosmic outcomes. Of course, science can safely predict—on the basis of how the cosmos has obeyed recurrent physical routines in the past—that the "laws" of nature will not be violated in future movements of stars, planets, microbes, and human societies. Yet science can no more predict what meanings may accumulate in the cosmic story, as it moves along, than your knowledge of the rules of grammar can predict what I will be saying later on in this chapter. Science is good at uncovering the skeletal loom of "regularities" into which cosmic process is woven, but it cannot grasp any hidden narrative meaning that may be stitching itself into the invariant physical principles and operations.

Seduced by the idea that nature is driven physically by mindless "laws," the archaeonomic reading has led to the widespread but dubious conviction that nature is meaningless. Simultaneously, archaeonomy's "grammatical" fixation on nature's so-called laws has engendered the unsupportable idea that human freedom is an illusion. This unfortunate development in human thought testifies to the seductive power that metaphors can play in shaping scientific thought —and not just religious understanding—sometimes for centuries.

Cosmic pessimists have usually mistaken nature's invariant regulations for deterministic physical forces. Blind to the dramatic character of world process, their archaeonomy has confused physical or chemical constants with mechanical forces mindlessly shoving the world toward a destiny fully determined by the past history of matter. By thinking literally of nature's inviolable habits along the lines of the juridical metaphor of lawfulness, archaeonomy not only has failed to notice the narrative insideness of the universe but it has also distracted several generations of scientists and philosophers from acknowledging and examining the obvious fact of their own subjectivity, their inviolable freedom, and their capacity to respond to rightness.

How the Universe Lost Its Mind

How did all this happen? Not surprisingly, the contemporary expulsion of cosmic purpose has occurred in collusion with the archaeonomic expunging of subjectivity from the physical record. To get rid of serious intellectual interest in cosmic meaning, scientific naturalists first had to remove all traces of insideness from nature as such. Earlier in human history the pervasive throbbing of interiority in the natural world seemed obvious. Before the rise of science, all of reality, not just plants and animals, often struck people as being alive and subjectively awake. Sky, sun, stars, rivers, rain, mountains, and rocks all pulsed with an inner life. Life was real and death unreal. Primal peoples therefore wondered how anything could be dead if everything is alive. If life is the norm, they assumed, then death is an unintelligible exception, indeed, an illusion. Consequently, to save their worldview in face of the fact that living organisms die, our "panvitalist" (all-is-alive) religious ancestors spontaneously devised the idea of a spiritual world in which the subjective center of experience that animates every organism still lives on after mortal life has drained away.[3]

The idea of a spiritual world where the subjective centers of living organisms can survive death is still the source of meaning and consolation for countless people on our planet. The prospect of immortality gives purpose to their lives. Archaeonomic naturalism, however, has no place for it. The universe, it claims, is a purely physical set of phenomena and, in the final analysis, mindless and lifeless.

After Descartes (1596–1650) the idea of mindless matter became the intellectual foundation of much modern scientific and philosophical thought, and it continues to influence intellectual culture.[4] A fundamentally mindless universe (Descartes's *res extensa*) still has an irresistible intellectual appeal for the simple reason that dead matter can be weighed and measured, whereas vitality, sentience, and other qualities associated with interiority cannot. Aliveness and consciousness, in that case, can easily be thought of as unreal.

Modern cosmic pessimism finds a comfortable home in the intellectual setting of an essentially lifeless and mindless universe. Subjectivity and hidden meaning are not allowed to exist in this kind of world. From the assumption that life is unreal it is easy to jump to the conclusion that subjectivity is not real either. The disavowal of subjectivity, in turn, has made it nearly impossible for contemporary thinkers, including exponents of Big History, to entertain the idea that the cosmos harbors an inside story, one that might carry an imperishable meaning. Accordingly, the religious search for meaning can easily be interpreted as pure escapism, an unwillingness to face the pointlessness of the real world.[5]

Unfortunately, the analogical reading of the universe, at least in modern times, has made its own kind of peace with the idea of an essentially lifeless physical universe. Analogical spirituality brings the comforting news that our hidden subjectivity can be saved, but only by supposing that we do not really belong to the physical universe. The physical world may serve as a place of moral discipline or as a point of departure for the spiritual pilgrimage, and the fragile beauties of nature may even function as symbolic pointers to a better world beyond, but analogy no less than archaeonomy is mostly blind to any possible narrative meaning still working itself out in nature.

The anticipatory vision, on the other hand, provides a fruitful new framework for rehabilitating both subjectivity and the religious sense of cosmic purpose. The "purpose" of the universe, in this reading, is that of awakening—especially by way of life, mind, and religion—to rightness. And the purpose of our own personal lives is to augment this awakening. Subjectivity, then, *is* the universe in the state of opening itself to the implantation of meaning. And religion is the universe in the state of grateful awakening to the dawning of rightness.

Making a case for the intellectual plausibility of such a view of the cosmos and human purpose, of course, requires that we critically examine, even more carefully than before, the creed of archaeonomic materialism according to which the most real and intelligible state of being is that of lifelessness. Materialism, according to Hans Jonas, amounts to nothing less than an "ontology of death." It is a world-view based on the belief that the most real and intelligible state of being is that of lifelessness.[6] No doubt, the ontology of death may appear at first to have ample empirical support. Life, after all, arose only after billions of years had passed during which the cosmos carried no animate organisms, and the thin spatial distribution of life even here and now seems to corroborate the archaeonomic impression of an essentially lifeless universe. Since life is so sparingly distributed spatially, and since it has made its debut so late in cosmic history, can we deny that lifeless "matter" is the metaphysical foundation of everything else, including our minds?

A universe that barely puts up with the fleeting existence of life and mind can easily seem to be essentially lifeless, mindless—and meaningless. Anticipation, however, denies that either science or reason compels us to think of the universe in this way. On the contrary, by interpreting the universe narratively, anticipation can read the early inorganic universe not as dead but as preliving. The universe, from a dramatic point of view, is not really indifferent to life and mind, and it never has been. Even spatially, life may not necessarily be as thinly distributed and terrestrially bound as we used to think. Astrobiologists are now speculating with justifiable confidence that life may be present at many locations in the cosmos. Recent discoveries of "extremophiles"—strange organisms that thrive in seemingly hostile places on Earth—suggest that life could be erupting bountifully at other apparently unpromising sites in the larger universe. Additionally, the detection by astronomers of a surprisingly high percentage of carbon compounds in interstellar dust enhances the probability of life bursting out wherever the appropriate environmental conditions are available.[7]

Even more important, however, astrophysics has lately been replenishing the world with life across the long reaches of time. Narratively speaking, as cosmologists are now coming to realize pursuant to new discoveries in astrophysics, the universe was never lifeless and mindless. Preparation for the long dramatic adventure of emerging complexity and the eventual arrival of life was already going on in the earliest moments of cosmic history. Even if it turns out that Earth is life's only host, we no longer need to consider our universe to be nearly so grudging toward the advent of life as pure materialists previously supposed. Life, along with its eventual transformation into different modes and degrees of subjectivity, when seen from a contemporary astrophysical perspective, is not a local anomaly but a property of the whole cosmic story.[8] So also, as I have been suggesting, is religion.

When I first started teaching courses in science and religion at Georgetown University in the early 1970s, most philosophers and scientists espoused, at least unconsciously, what Jonas calls an ontology of death. They took for granted that matter is at best indifferent, if not downright hostile, to life. One of the texts I assigned to my students was Jacques Monod's *Chance and Necessity*, a celebrated work by a Nobel Prize–winning biochemist whose main message was that life and mind have no business being here—or anywhere. Monod's universe was fundamentally lifeless and mindless. Inventing a curious alloy of atheistic existentialism and materialist mechanism, Monod took for granted that the only reasonable way to understand nature is with the absurdist postulate that life and mind are physically improbable accidents and that true awakening means coming to recognize that the universe, objectively speaking, is pointless.[9]

The cosmos, however, no longer looks quite the same as it did to Monod. Sooner or later, scientific storytellers now increasingly assure us, the emergence of life in the Big Bang universe would have occurred inevitably. Yet archaeonomic materialism, somewhat chastened perhaps, still has a stranglehold on intellectual life. Though less secure in their assumptions than Monod, its disciples still cling to an only slightly softer version of the modern ontology of death. In spite of their reluctant admission that the Big Bang universe is "just right" for life and mind, influential intellectuals continue to exile subjectivity and purpose from the sphere of true being. A universe that has lost its mind still seems intellectually preferable.[10]

The astronomer Martin Rees, for example, argues that even though our Big Bang universe happens to be just right for life and mind, in all probability a larger—and pervasively lifeless—abundance of "universes" loiters invisibly in the background. Outside the range of empirical investigation, he speculates, there exists an unimaginably larger and lifeless manifold of worlds, a statistical largesse that permits at least one life-bearing cosmos to pop up unexpectedly.[11] The larger assembly of universes, according to this schema, remains indifferent to life and mind, so nature is still fundamentally dead. The possible existence of a lifeless multiverse apparently washes out whatever dramatic importance our own anomalous life-bearing Big Bang experiment may seem to have.

Materialist cosmologists assume that if the absolute number of universes is immense and if opportunities for blind experimentation in them are sufficiently generous, the laws of probability allow that a life-bearing universe may turn up sooner or later in the lottery of worlds without our having to revise the assumption that the natural world is dead and pointless at heart. Life and mind in that scenario would still be outcomes, on a larger scale, of a purely accidental play of numbers combined with the rigor mortis of implacable physical laws. To the archaeonomic naturalist, then, the awakening that takes place in the emergence of life and mind on our planet need not overturn the prevalent pessimistic understanding of nature. When life and mind do show up—perhaps in only one (or maybe even several) of the countless universes—their existence does nothing to subvert the learned assumption that the cosmos has no overall meaning.[12]

Let us suppose, then, that a multiverse exists. After all, some contemporary cosmologists, independently of our considerations here, consider the idea of many universes to be theoretically compatible with the notion of a very rapidly expanding early Big Bang universe.[13] Yet even if a multiverse exists, does this mean that the collective multiplicity of worlds is fundamentally lifeless and mindless? Is our own life-bearing universe an aberration within a pervasively lifeless multiverse?

The materialist way of understanding the world as represented by Monod, I suggest, is the product not of scientific inquiry but of the "archaeonomic illusion." This is the fallacy that we can render all cosmic outcomes fully intelligible by excavating, in ever more minute detail, the past stages of their emergence. The archaeonomic illusion equates the quest for understanding with reductive simplification, a dubious way of looking for intelligibility. We may also call it the "analytical illusion," since it assumes that breaking complex things down into their elemental components is sufficient for their explanation.[14] An analytical method of exploration is essential to science, of course, but what I have been referring to as archaeonomic explanation is not science. Rather, it is a scientifically indemonstrable belief. It is not self-evident, after all, that the only way to get a handle on complex cosmic outcomes is by reducing them to their component parts or by uncovering their earlier and simpler physical past. The archaeonomic reading, rather than waiting for coherence to emerge out of the world's elemental particulars, unreasonably makes past atomic de-coherence the "fundamental" state of cosmic existence. Rather than scanning the horizon for a still-emerging narrative coherence that alone can satisfy the human hunger for understanding, pure analysis leaves our minds stranded in a primordial desert of scattered physical units, a dissolute state of being that the cosmic journey itself left behind ages ago.

The anticipatory vision of nature cannot settle for such a shrunken idea of being and intelligibility. In the universe that our minds actually inhabit, to find true being and meaning we must wait, actively and attentively, for the dawning of deeper narrative coherence. Again, there is nothing objectionable about the analytical method as such, especially since it is essential to our learning many details of the story of the universe. What is questionable, however, is the assumption that reductive analysis by itself can satisfy our longing for intelligibility. Our search for understanding cannot be assuaged, it is worth repeating, only by breaking present complex entities down into their earlier and simpler subordinate parts. An exclusively reductive look into the cosmic past, important though it may be as part of a larger movement of the scientific mind, is not enough to slake our thirst to understand what is really going on in the universe. If we want to make sense of a universe that is still coming into being, it is not enough to look at how elemental physical stuff has been manipulated by inviolable laws and past efficient causes. We need, even more, to look for a narrative coherence that can be dawning only farther along in the story, an intelligibility that currently lies mostly out of our grasp. The quest for cosmic purpose, it follows, cannot be undertaken reasonably apart from an anticipatory thrust of our minds in the direction of the not-yet.

Making Sense of a Multiverse

Anticipation's deliberate discipline of patient expectation may even allow us to make sense of the idea of a multiverse. Instead of seeking to "explain" the physical properties of one lonely life-bearing universe—our Big Bang cosmos—by statistically multiplying dead ones, as Rees and others do, let us make our present universe the point of departure for exploring the narrative meaning of a (theoretically possible) multiverse out of which our own may have emerged. Imagine a journey back archaeonomically into the primordial cloud of lifeless universes from which our life-bearing universe now stands out. Then, having traveled mentally back into that collective multiplicity of unseen worlds, let us reverse our perspective, turn around half-circle, and look toward what may still be coming to birth out of that hypothetical statistical immensity. Instead of trying to "explain" our life-bearing, Big Bang universe by assuming an immense plurality of lifeless ones—an approach that can only lead our minds back into the deadness of de-coherence—let us entertain the thought that the eventual emergence of our life-bearing universe brings an unexpected narrative coherence to the multiverse from which it has been born.

In reading any story, we find its meaning or coherence not only by looking back into an original plurality of scattered components or episodes, but by looking up ahead toward the syntheses that may be emerging from that earlier plurality.[15] Can we really understand atoms, for example, before we wait and see, after their initial emergence, what they can do later on in the cosmic journey when they link up with other atoms to form organic molecules? Or can we understand earlier macromolecules until we wait and see what they can do when combined (later on) into living cells? Finding intelligibility in our universe, at least from an anticipatory perspective, means fitting things narratively into larger wholes, not breaking them down into their elemental constituents. Once the universe is understood as a story still in progress, therefore, we may expect its component moments and elements to become intelligible only if we wait and see how they fit into a story still being told. And if the story is not yet over, intelligibility, meaning, and purpose are qualities we must not be too impatient to possess.

It follows that if a multiverse exists, it cannot by itself be taken as explanatory of our distinctively life-bearing universe, as archaeo-nomically inclined multiverse advocates speculate. Just the opposite: The existence of our one life-bearing universe may be thought of as bringing intelligibility, or narrative coherence, to the many preexisting worlds from which it is now coming into being. According to an anticipatory reading, in other words, the intelligibility of a multiverse may emerge only as it moves into the territory of the not-yet.[16] Moreover, even if there exist many lifeless worlds now hidden from view, they too may still be narratively complicit in the emergence of life in our own. The idea of a multiverse therefore cannot reasonably be used to shore up scientific materialism and cosmic pessimism. Dramatically speaking, in any case, nature is *one* story, even if its many subdivisions or tributaries remain empirically isolated from one another. Even if most of the hypothesized worlds in a multiverse are locally devoid of living and thinking beings, and even if the quality of time, space, and the laws of nature in these worlds is radically different from our own, they are nonetheless essential aspects of a larger, single, still-unfolding story of the emergence of life and thought.

Multiverse theory, then, regardless of the suspect motivation that makes it attractive to cosmic pessimists, need not be thought of as supporting the modern ontology of death. No matter how many physically separated "universes" there may be, they would still enjoy a narrative togetherness inasmuch as they all participate in the wider drama of the arrival of life, mind, and religious restlessness—even if the latter outcomes actually show up only in our Big Bang cosmos and, within that setting, solely in its terrestrial quarters.

An anticipatory vision, in summary, wagers that what is most real, important, and intelligible can be approached only by expectation, not by retrospection. Neither our own universe nor a possible multiverse would ever, therefore, have to be thought of as essentially dead and dumb. Life on Earth, in this reading, need not be considered epiphenomenal, nor the universe purposeless. In an unfinished universe (or multiverse) a disposition of expectation is epistemologically essential to the opening of our minds to its intelligibility. Purpose, then, can make its appearance only uncertainly on the horizon of the not-yet. We would not grasp it as much as it would grasp us.

Purpose as Cosmic Aim toward Beauty

Purpose, I have been saying, means the realizing of rightness. The universe has a purpose insofar as it is awakening to rightness, and religion yokes human life and consciousness to that wider awakening. The universe (or, if you prefer, the multiverse), however, is still coming into being and remains open to richer syntheses in the future. This is why we cannot realistically expect it to be fully intelligible at present. All the more, then, in our quest for cosmic purpose, do we need to look steadily toward the not-yet, following the discipline of patience. Religion, even in all of its ambiguity, is a stimulus to the practice of that discipline.

Unless the religious affirmation of the reality of indestructible rightness is true, of course, a perishable universe would have no point or purpose. Religion is cosmically significant because of its confident belief that there does indeed exist an eternal ground of meaning. The notions of *God, Allah, logos, wisdom, Brahman, nirvana, moksha, dao,* and *kingdom of heaven* all express the religious intuition of an indestructible rightness that saves the world from final insignificance. Hidden from scientific objectification, a transcendent dimension of imperishable rightness keeps the whole inside story of the universe from being completely dissolved in the flux of its external temporal passage. Affirming fully the truth of religion, however, can take place only by simultaneously embracing the discipline of patience that is central to what I have been calling faith.

An anticipatory perspective adds to all of this that both the universe and human existence, in order to have meaning, must be an endless quest. Terminate the quest and you remove the sense of meaning. A moving horizon of indestructible rightness gives meaning to all passing things, but it remains forever an elusive ideal and not a final possession. This is why it is the height of silliness when celebrated archaeonomic naturalists demand that the cosmic purpose anticipated by religion make itself immediately available or "evident" to the objectifying gaze of science.[17] By definition, the indestructible rightness that gives purpose to all that passes cannot be one among other transient objects available to immediate experience or scientific objectification.

From the beginning of this book, however, I have associated

rightness not only with infinite being, meaning, goodness, and truth, but also with infinite beauty.[18] Perhaps, above all else, it is its anticipation of limitless beauty that gives the universe an overarching purpose. Beauty is the harmony of contrasts, the ordering of novelty, the unifying of multiplicity, the gathering of diversity into deeper relationship. This, at heart, may be what the universe is all about. I believe that in an unfinished universe, without the slightest conflict with science, people of many faiths might now agree that the ultimate aim of everything is beauty.[19]

There is nothing, it has been said, that cannot be redeemed by beauty.[20] Beauty is the greatest of present facts but also something waiting to be realized. The universe in that case is purposeful because it is an adventurous aim toward the endless intensification of beauty.[21] To anticipation, the aim toward aesthetic intensity is the central theme of the cosmic story, and subjectivity is the most intense concentration of the cosmic aim toward beauty. Religion implies that the "secret essence" of the universe is being "garnered" by an ever-expanding beauty that in some sense is still waiting to be realized.[22] From an anticipatory religious point of view all cosmic events, including the whole evolutionary story of life and the struggles and triumphs of human history, with all its losses and gains, may be redeemed by being woven into an everlasting beauty, a goal that can reveal itself symbolically here and now only fragmentarily and fleetingly.[23]

This "aesthetic" way of understanding cosmic purpose should lead us to react with unprecedented moral and religious horror at the modern human destruction of Earth's precious and precarious ecosystems. Instead of cultivating the ideal of beauty as the aim of all things, we on Earth are now turning away, in what amounts to despair, from the anticipatory sense that beauty is still "waiting" to be realized. An aversion to aesthetic rightness is evident in current environmental abuse that, in effect, is reducing Earth's biodiversity to the monotony of elemental lifelessness. Unfortunately, the academically sponsored archaeonomic reading of the universe provides tacit intellectual justification of this reduction. By claiming in effect that life and mind are not really more than the lifeless stuff from which they have come, archaeonomic naturalism, with its implicit ontology of death, in principle provides no reasonable support for a robust ecological morality intent on preserving living complexity.

Here again, however, we encounter the delicious irony of archaeonomic naturalists not really believing what their worldview logically implies. Earlier we noted that strict scientific materialists deny that subjectivity actually exists while at the same time illogically attributing exceptional value to their own intellectual interiority. True to form, with respect to one of the most troubling issues of our time, some of the most ecologically sensitive people around—I am thinking, for example, of E. O. Wilson and his many admirers—are also deeply committed archaeonomic naturalists. On the one hand they are willing to make great personal sacrifices to save and nurture terrestrial life. On the other hand, their materialist metaphysics logically removes any permanent reason for doing so.[24] Analogy, by contrast, emphasizes the sacramental value of life and interprets our ecological destructiveness as a kind of sacrilege. By diminishing life on Earth, it maintains, we weaken the religious sense of nature's present participation in eternity.[25] Once again, however, analogy's sacramental sense of the world typically fails to take into full account nature's narrative constitution and its openness to the arrival of more-being.

From the perspective of anticipation our current ecological recklessness amounts to something much more dramatic than either archaeonomy or analogy is able to see or feel. Since an unfinished universe has room to blossom forth into new and indeterminate forms of beauty in the future, our present ecological neglect amounts to an estranging of the universe from its ultimate destiny.[26] By destroying the beauties of nature here and now we are in effect suppressing the anticipatory orientation of a whole universe. Anticipation, I believe, is a richer metaphysical setting for ecological responsibility than either archaeonomy or analogy. Anticipation opens the natural world, and not just human souls, to the enhancement of beauty far into the indefinite cosmic future. Ecological responsibility, therefore, has to do not only with preservation of past cosmic achievements and terrestrial ecosystems, but also with preparation for the augmentation of life and the emergence of more-being in the liberating expansiveness of what is not-yet.

CHAPTER NINE

Obligation

Conduct is a by-product of religion—an inevitable by-product, but not the main point. Every great religious teacher has revolted against the presentation of religion as a mere sanction of rules of conduct. . . . The insistence upon rules of conduct marks the ebb of religious fervor.

—ALFRED NORTH WHITEHEAD

PEOPLE COMMONLY ASSOCIATE RELIGION with morality and lists of commandments. The philosopher Immanuel Kant (1724–1804), however, argued that religion is not the source of morality. The moral law is already stamped silently on our hearts by birth and needs only to become explicit in the course of our lives. It is not the main job of religion to make us behave. Religion can give us portraits of exceptional individuals whose lives teach us how to obey the moral law within, but religion is much more than a list of do's and don'ts.[1]

Jesus, for example, is not the sole author of a new ethical imperative to love our enemies. Rather, he is a man who obeys in an exemplary way the universal commandment to love foe and friend alike. The imperative to love one another is already inside each of us, but it needs to be brought to the surface, and this is where religion comes in. Meditating on stories about Christ or Krishna can activate in us a willingness to do our duty, but the commandment to love is already sown deeply into our souls. All humans, Kant tells us, are endowed with an interior sense of "oughtness," but we need exemplars to teach us how to shape our lives in accordance with the internal call to goodness. We misunderstand religion, then, if we take it to be primarily a source of ethical instructions. Religion is not so much a guide to conduct as an answer to the question "Why should we bother to be good at all?"

It is hard, nevertheless, to distinguish clearly the moral call to goodness from the religious attraction to "rightness." Biblical religion maintains that worship of God includes caring for the needy. True religion means visiting widows and orphans, says the New Testament (James 1:27). Islam, Hinduism, Buddhism, and other traditions all take good works to be part of religious life. Yet being religious is not necessary for people to be good. Nonreligious people can be highly moral, and religious people can be flagrantly immoral. So far I have associated religion with an awakening to rightness, but awakening to rightness can be intellectual, aesthetic, and moral without necessarily being religious. What, then, does religion have to do with morality, if anything?

We can answer this question in a fresh way today if we take into account the unfinished status of our universe and the fact that each of us is part of a cosmic birth process. Our actions, in this new setting, are moral to the extent that they contribute to the process of cosmic awakening and transformation. Morality is more than this, no doubt, but here I want to consider what it means in terms of the new scientific impression that the cosmos, and everything in it, is somehow not-yet. Our new dynamic cosmic perspective makes a difference in how we understand not only religion and faith but also our moral lives. To clarify this difference let us look comparatively at what the human sense of obligation means in terms of the archaeonomic, analogical, and anticipatory ways of reading the universe.[2]

Archaeonomy. As expected, archaeonomic naturalism tries to explain the human "moral instinct" by tracing its origin and development back into the past history of life and the physical universe. Archaeonomy agrees with Kant that religion is not the direct source of moral norms, but it rejects Kant's notion that moral imperatives somehow come from beyond. Religious instruction is not needed to plant a sense of duty in our hearts. We are endowed with moral instincts because of our having evolved like all other organisms by way of an impersonal process of natural selection. To survive in a difficult world, all living organisms not only have to compete but also cooperate with others, and human morality is rooted in this biological imperative. Without the cooperative support of other members of a species the individual organism cannot survive long enough to bear offspring. Evolutionary fitness, according to contemporary evolutionary biology, includes being born and raised in a life-context where the general population of genes in one's group or species fosters some degree of cooperation. The evolutionary naturalist theorizes that we are moral beings, therefore, not because of a divine command given with our existence, or because human conscience participates in an infinite goodness, but because of the biological need for cooperation among members of our species.

In this account we are moral because in the past history of life cooperative behavior has turned out to be more adaptive than uncooperative behavior. Only by going back—archaeonomically—into life's evolutionary past, then, can we learn how the sense of obligation came to be such an important part of human existence. As recent versions of the Darwinian story tell it, an exceptionally strong imperative to cooperate arose by a series of biological accidents in prehuman evolution. The emergence of modern humans as a distinct and dominant species depended on our remote biological ancestors eventually becoming unintentionally endowed with cooperative sets of genes that enabled them to conquer major threats to their existence. It was not God but a combination of random genetic events and natural selection that gave rise to cooperative human tendencies. Cooperative instincts were then passed on through human history by genetic rules of inheritance all the way to the present, and this biological transmission accounts for why we have moral instincts today. The moral law, therefore, is not really inside us but outside us. It is embedded in a scientifically specifiable, purely physical flow of genes that carries us along and makes us behave well enough to survive in an indifferent universe.[3]

Religion, one can see, does not have a central explanatory role in this account of morality. Defenders of analogy and traditional religion, of course, protest. How else than by divine inspiration and participation in an eternal goodness can we account for the most heroic instances of moral behavior—for example, when humans lay down their lives for others? Can evolution, in other words, give us the right explanation of all our altruistic tendencies? Yes, the evolutionary moralist replies. Altruism is the instinct that leads an organism to sacrifice its own genetic future for the sake of the survival of the whole population of genes it shares with its kin. Altruism, in this sense, is not an exclusively human trait. A prairie dog, for example, can sacrifice its own life for the sake of others. In the act of warning its colony that a hawk is nearby, it may stick its neck too far out of its hole and get eaten up in the process. That particular member of the community may not survive, but in the act of self-sacrifice its altruism contributes to the survival of a whole pool of kindred prairie dog genes.[4]

Altruism in a broadly genetic sense, then, was already part of the evolution of life long before humans came along. The sense of oughtness—which to the evolutionary naturalist entails altruism as defined above—was many millions of years in the making. It did not slip silently into each human soul at a special moment of divine creation. It is a purely natural development that scientific excavation can now fully unearth. We can neither understand nor appreciate our own sense of obligation, therefore, apart from biological accounts of morality's gradual emergence in reptilian, mammalian, primate, and eventually human evolution. To arrive at a complete and accurate portrayal of how our own moral sense arose, we need to journey back into the earliest, previously obscure chapters in the history of life. Long before humans came along, we now understand, social insects like ants and bees were already implementing evolution's need for cooperation as a necessary adaptation. Had social organisms not learned to cooperate with one another, they would have died out. We humans are moral, then, because our animal and humanlike ancestors chanced upon cooperation as a survivable trait and passed it on to us. We call it morality, but it is nothing to boast about.

Since archaeonomic explanation means accounting for present phenomena by tunneling back into the cosmic past, its ideal of intelligibility is that of retrieving a chain of physical connections starting from as long ago as possible. In the case of our sense of obligation, a series of aimless efficient causes led physically, step by tiny step, to the intriguing evolutionary outcome we call human morality. Here, once again, the archaeonomic reading of nature rests its understanding of everything going on now—and everything that will happen in the future—on what has happened already. Human moral instincts, therefore, do not exist because of the world's awakening to rightness, as an anticipatory account maintains. Our intuition of rightness and wrongness is not a response to the self-revelation of an infinite goodness, as analogy believes. Rather, our sense of obligation is a set of instincts deterministically deposited in human genomes and organisms by material causes originating ultimately in the remote cosmic past. Archaeonomic naturalists sometimes allow that human cultures can sculpt our moral sensitivities in unique ways, but the fundamental explanation of the human sense of obligation is ultimately a physical one.

Far from rooting morality in religion, moreover, the archaeonomic point of view sometimes even allows that morality is the source of religion. Evolution first crafted humans to be cooperative, Darwinian moralists theorize, but early in human history religious illusions arose to shore up cooperative behavior. Our ancestors chanced upon the notion that they would live more congenially together if they believed in a system of supernatural rewards and punishments. Good behavior is more likely to take hold in a population of humans if a majority of them are tricked into believing that they will be rewarded for it, and if simultaneously they are convinced that uncooperative behavior will bring punishment both during life and after death. Religion with its belief in immortality arose to reinforce cooperative behavior by giving people spiritual motives to behave rightly. Ideas about deities, and eventually fantasies about the existence of a supreme being, originated in the need for a system of supernaturally administered rewards and punishments. In this version of archaeonomic understanding, religion, accompanied by ornate assemblies of deities and florid mythic fantasies, arose out of morality, not vice versa.

In accounting more specifically for how religion came about to support moral life, evolutionary naturalists fill in the narrative gaps in elaborate ways. Long before humans appeared, according to one recent auxiliary account, evolution had already equipped living beings with skills for escaping predators. Our humanlike ancestors, beginning several million years ago, had to develop the talent for looking out for beasts of prey that might threaten their existence. Without such a faculty they would not have survived and borne offspring, and we would not be here. Physiologically the prowess for predator detection required the evolutionary creation of complex brains and nervous systems that can imagine the existence of agents of harm even when the latter are out of sight. Then, as a by-product of the ability to imagine unseen perpetrators of mischief, ancestral brains could visualize the existence of all kinds of other unseen agents, including deities who may reward and punish. The capacity for imagining such fictions could in turn prop up morality and thus enhance the probability that humans would cooperate, survive, and then reproduce in sufficient numbers to make us the dominant species on Earth.[5]

So goes a popular current evolutionary narrative of the conditions essential for morality to function adaptively. It deserves our consideration, and I am willing to follow the Darwinian naturalists as far into the past story of life as they want to go. Certainly morality has an adaptive evolutionary function in some—though not all—environments. Yet this adaptive function does not exhaust its meaning. An anticipatory reading of nature, as I will argue shortly, allows us to envision morality as more than adaptation, and religion as more than an evolutionary shoring up of morality. Moreover, in an unfinished universe morality, like religion, has considerable room for development and maturation. Morality is an important way in which human consciousness gradually awakens to rightness, and so there can be distinct stages of moral development, a point that evolutionists nearly always overlook. Even in its infancy and immaturity, however, morality has a meaning that remains unexplored as long as we fail to situate it in the context of a universe that has yet to be fully actualized. More on this below.

Did Morality Give Rise to Religion?

Before looking at the analogical and anticipatory readings of morality, I need to point out that Darwinian theorists are not alone among scientific naturalists in claiming that the sense of obligation gave rise to religion rather than the other way around. In the twentieth century, for example, the philosopher Max Otto (1876–1968) theorized that moral aspiration has existed much longer than ideas about the gods, and that it will persist long after all the gods are gone. For even in the absence of religion there is "devotion, heroism, self-sacrifice, loyalty to causes." The natural human attraction to virtue, Otto allows, sometimes gives rise to noble ideas of God, but moral greatness "has been achieved by individuals and by whole peoples in the absence of faith in God."[6] More recently, the philosopher Kai Nielsen has argued that human moral sensitivity is the source of the idea of God. People cannot sincerely engage in acts of worship, after all, unless their God is conceived of as an exemplification of the moral rightness they already treasure. "A religious belief depends for its viability on our sense of good and bad—our own sense of worth—and not vice versa." "A moral understanding," Nielsen goes on to say, "must be logically prior to any religious assent."[7]

Other archaeonomic moralists go even farther than Otto, Nielsen, and present-day evolutionists in positing the priority of moral rightness over religion. For them religion is not only morally unnecessary but also intolerable. The nineteenth-century European philosopher Friedrich Nietzsche, for example, in his fascinating genealogy of morals, speculated that religion, in particular Christianity, is the expression of a defeatist attitude toward life. In his professed obedience to a rightness deeper than that idealized by religion, Nietzsche proclaimed the "death of God" as the greatest event in human history. Removing the looming impression of God from human consciousness is right, he thought, since it will clear the way for a "higher," religionless morality, a nobler set of imperatives that can replace the authoritarian sense of obligation instilled in people by conventional piety. After the death of God, a superior version of humanity would place morality on an entirely this-worldly foundation. Where people are fully alive, according to Nietzsche, they do not need to ask whether and how to be moral. The sense of obligation will arise naturally and spontaneously. A higher morality, he assumed, would emerge only after religious faith is gone.[8]

Prior to contemporary evolutionary accounts many other critics had also claimed that religion is not essential to morality. Karl Marx acknowledged that religion is an understandable response to human suffering, but he declared that the promise of rewards in the "next world" has diverted moral life from its real obligation to overthrow the oppressive socioeconomic conditions that give rise to religious escapism. Marx's understanding of obligation was archaeonomic too, in the sense that he looked back into natural history, materialistically understood, and into the history of economics, social arrangements, and class conflict, to expose the purely natural reasons why religion persists, and why human ethical life can now be better off without it. Sigmund Freud, another archaeonomic thinker, dug down into what he called the unconscious, a submerged region of the human psyche where irrational tendencies, the libidinal source of both sexual and religious instincts, had accumulated during the course of natural and human history. According to Freud, psychoanalysis can remove the clutter in instinctual life so that morality may now be placed on a purely rational foundation.[9]

The French writer Albert Camus, through the character of Dr. Rieux in *The Plague*, gave voice to another, and more common, moral criticism of religion. Religious references to an all-knowing, omnipotent being who permits suffering has the effect of legitimating our own indifference to the suffering of other living beings, a flaw that makes God-consciousness morally offensive. "Since the order of the world is shaped by death, mightn't it be better for God if we refuse to believe in Him and struggle with all our might against death, without raising our eyes toward the heaven where he sits in silence."[10] More recently the New Atheists mentioned earlier have justified their passionate rejection of religion by cataloging instances where religious "faith" sanctions violence and oppressive behavior. In the name of moral rightness, they too argue, religion must go away, once and for all.

Finally, the French existentialist philosopher Jean-Paul Sartre, more forcefully than any other modern or contemporary philosopher, insisted that morality is invented not by God but by human beings. Belief in divine commandments is an excuse for not facing the fact that each person, and no one else, is responsible for deciding what is right and wrong. Rightness, in the Sartrean perspective, cannot awaken us since we are its creators. We awaken rightness, as it were, and along with it we invent whatever meanings we decide to give our lives. It is always and unconditionally right, Sartre taught, to deny that there is any transcendent source of rightness![11]

Analogy. Moral values, according to analogy, are grounded in an eternal goodness. They are not our own creations but God's. It is the business of religion to point us toward the invisible, indestructible rightness that accounts ultimately for why humans have moral inclinations. The voice of conscience is a sign of the human soul's inescapable participation in an eternal source and standard of goodness. In the analogical reading, therefore, rightness is eminently real. It is not a personal concoction, a cultural fiction, or an evolutionary invention. So the solution to modern social, political—and ecological—ills is to reconnect our hearts and minds to the eternal moral order now partly lost. It is religion's function to help us do so.

Analogy, furthermore, claims to have found a serious contradiction at the very heart of all naturalistic attempts to divest morality of its connection to religion. The problem is that in the very act of trying to cleanse the world of any religious motivation for morality, archaeonomic naturalists are drawn by a tacit assumption that it is *right* to do so. Analogy, therefore, wants to know how the silent appeal to an a priori unconditional rightness by secular moralists could be justified if the sense of right and wrong were exclusively a human invention or an evolutionary adaptation. If moral imperatives were nothing more than human fabrications or evolutionary creations, why should anyone be expected to follow them no matter what? After all, we might be driven by irrational instincts, and evolution can lead organisms to do some very cruel things.

The transcendence and a priori reality of rightness, analogy insists, does not go away for all our attempts to deny it. It trails us in the back door even while we are sweeping it out the front. In the act of clearing out the annoying undergrowth of religion to make room for a superior, purely natural morality, evolutionary moralists witness strangely to an irremovable rightness without which their own moral advice has no reasonable justification. In claiming that the rightness they profess to have found is nothing more than a result of random variations and natural selection, they have destroyed any good reason why their readers would have to take them seriously, either morally or intellectually.[12]

In the three-way conversation we have been following, then, the special significance of the analogical vision is its proper sense of the logical incoherence of any purely archaeonomic interpretation of the human sense of rightness. Analogy correctly insists that moral imperatives are empty unless they are connected to an indestructible rightness. For a very good reason, then, the great religious and philosophical traditions have always pointed to a timeless order of "rightness" as the basis for a realistic and rationally justifiable moral consensus. Egyptian religion calls this eternal order *ma'at*. Abrahamic faith names it the *wisdom*, the *word*, or the *will of God*. Daoism refers to it as the *dao*, or the *way*. Platonic philosophy locates the source of moral obligation in an eternal *goodness*. Hindus submit to *rta*, Buddhists to *dharma*, Confucians to *jen*, *li*, and *yi*. Other traditions may have alternative names for the ground of moral rightness, but religion everywhere roots morality in a timeless source and standard of values that we humans did not invent.

If no everlasting criterion of rightness and wrongness exists, the novelist Fyodor Dostoevsky (1821–81) wonders, what is to keep each of us from deciding on our own—and arbitrarily—what is good or evil? A tormented character in *The Brothers Karamazov* provides a classic example of the passionate analogical protest against naturalistic moral relativism: "If [God] doesn't exist, man is the chief of the earth, of the universe. Magnificent! Only how is he going to be good without God?"[13] In *Crime and Punishment* Dostoevsky asks why, in the absence of any eternal norm of rightness, it is wrong to chop open someone's head to steal a little pocket change. If rightness is relative to our own desires, especially since we are more irrational than rational, how will that keep the world from lapsing into moral chaos?

Numerous learned contemporary analogical thinkers have amplified Dostoevsky's apprehensions. Many of them, above all those attracted to classical theology and perennialism, observe that the most horrific crimes in human history have occurred simultaneously with the rise of archaeonomic naturalism. They wonder why other intellectuals and academics fail to see the connection between the secularization of conscience and the expulsion of subjectivity on the one hand and the ruthless slaughter of millions of people in the twentieth century on the other. If the ethical track record of religion is hardly clean, that of modern materialism, with its denial of subjectivity and eternal values, is much worse.

Rightness, analogy persists, cannot be wiped off the tablet of human consciousness as easily as archaeonomy supposes. No serious call for moral responsibility can account for rightness in an exclusively naturalistic or humanistic way without self-contradiction. Nietzsche and Sartre, for example, in spite of their explicit declarations that human beings are not bound by any eternal standard of rightness, implicitly witness to a hidden horizon of rightness that judges, condemns, or corrects their own moral pronouncements. Their explicit moral relativism is contradicted by the absoluteness of the moral imperatives they impose on the rest of us. Nietzsche, for example, tells us that after the death of God our real obligation is to remain uncompromisingly "faithful to the earth," but there is nothing in his moral philosophy or absurdist cosmology that tells us why this imperative is unconditionally right. And Sartre tells us that in the absence of religiously sanctioned moral principles, it is absolutely right to accept our freedom and wrong to flee from responsibility. Like Nietzsche, however, Sartre issues the imperative to live responsibly as having an absolute and unconditional claim upon us, but without telling us why. Even in the act of explicitly denying the indestructibility of rightness, the moral relativists cannot escape the dawning of rightness in their own lives, nor can they completely silence its awakening both them and us to authentic existence.[14]

Anticipation. If analogy is right to root morality in an incorruptible rightness, however, it is wrong in failing to connect our moral lives to a universe still coming to birth. Unaffected by new scientific discoveries, analogical thinkers do not bother to ask what it may mean for the moral life that we live in an unfinished universe. Archaeonomic naturalists, meanwhile, are aware that the cosmos is always in transition, but they see no moral relevance in the fact that it is still coming into being. Anticipation, however, maintains that an awakening universe gives moral life a whole new context and accent. It locates the immediate source of our sense of obligation neither in evolved instincts nor in a divine command planted immediately in our souls, but in our being borne along by a universe that is still awakening to rightness.[15] Here the moral life is motivated by the prospect of our contributing to the great unfinished cosmic work of letting in more-being and deeper beauty.

Anticipation can agree in part with both the archaeonomic and analogical understandings of morality. It accepts the new scientific awareness that morality is intimately connected with the cosmos to which we are all physically tied, and it also accepts the analogical intuition that rightness is real. However, what anticipation adds is that rightness, though real, still remains to be realized. Only the sense that something everlastingly right is dawning can awaken us morally, but only the fact that rightness is not yet fully actualized can give lasting significance to our moral acts and obligations.

Furthermore, anticipation allows that in a certain qualified sense religion is indeed morally superfluous, for it is undeniable that most nonreligious people are moral, often admirably so. Humans can have profound moral inclinations in the absence of any explicitly religious affiliation. Religion is not needed to give us moral codes and lists of imperatives. Yet a religious sense of rightness *is* essential if we are to address satisfactorily the question of why we should bother to be moral at all. Religion exhibits its distinctive importance not when it gets directly involved in the morality business—which of course it cannot avoid altogether—but when it gives its devotees reassurance that the universe they live in has a purpose and that the good life is one lived in conformity with that purpose.[16]

People today, however, do not generally have a strong sense of cosmic purpose, and both archaeonomy and analogy are partly to blame for this huge deficit. In different ways both foster distrust of the cosmos and discourage hope for its future. Their shared cosmic pessimism numbs moral incentive. An anticipatory reading of the universe, on the other hand, unlike its two rivals, offers assurance that, since the universe is still coming into being, it leaves room for the emergence of more-being, richer meaning, and more intense beauty—up ahead. Each of us can participate, at least in a modest way, in the ongoing creation of a universe that over the long haul has been undergoing an intensification of its being and beauty. Our moral lives presently, then, can gain new meaning once we acknowledge that we belong fully to a universe that has room to become even more beautiful, more alive, and more conscious than before.

Moral aspiration, that is, can thrive only where there is a prior conviction that life is worth living in the first place. Without that vitalizing premise a serious moral commitment to the pursuit of rightness cannot last across many generations. Whenever our lives seem to lack purpose, the will to live dwindles, and so also does the desire to do good. The primary role of religion, morally speaking, therefore, is not to formulate commandments but to motivate people to trust in the meaningfulness of the universe to which they are inseparably linked. Without a robust sense of cosmic purpose grounded in an indestructible rightness, the worthwhileness of our lives and the meaning of duty remain in question.[17]

It follows that the meaning of human love also shifts when, in our attempts to justify it, we move from archaeonomy through analogy to anticipation. To archaeonomy love of others is an adaptation built into us by evolution to get human genes from one generation into the next. How this evolutionary portrait of the moral life, if that is all there is to it, could ever motivate us to love one another and to do so across many generations has yet to be demonstrated. Learning from biology that our genes cannot survive without cooperation is hardly destined to promote either heroic self-sacrifice or even the everyday meeting of our obligations to one another.

Analogy professes to have found a reasonable justification of love in the unconditional value each person has by virtue of participating here and now in a timeless world of infinite goodness. This classic attempt to justify human love is certainly an improvement on evolutionary naturalism, but it fails to connect our works of charity to the newly discovered fact that the whole universe is still a work in progress. Analogy too easily settles formally for a kind of love that makes no lasting imprint on the cosmic course of events.

According to anticipation, on the other hand, we love one another not only for biological reasons, and not only because we each reflect an eternal goodness, but also because we are each part of a grand cosmic pilgrimage toward fuller-being, deeper subjectivity, and more intense beauty. Again, the imperative to love means much more than a cosmic perspective alone can assign to it. Yet the ethic of love might flourish more spontaneously if we could become fully aware, first, that the universe to which we all belong may have a future that has not yet been realized; second, that those we love have a future; and third, that their lives and destinies are tied inseparably to that of the universe and its future. Anticipation allows that our works of love not only respect the intrinsic value of each human person but also, in doing so, contribute something unprecedented to a universe still in the making. This cosmic motivation does not require that we necessarily cease what we are already doing, nor does it suggest that we can escape the humdrum duties of everyday life. It does, however, invite us to realize that in small and seemingly insignificant ways our moral actions, especially our works of love, may contribute something to a cosmic beauty still waiting to be realized.

Can We Trust the Universe?

An anticipatory perspective reads the universe as a currently incomplete journey toward fuller-being, and this impression can activate and energize our moral instincts in a way that was inconceivable to our ancestors. The prospect that the universe is still in the process of becoming more, and that beauty can grow, offers a previously unnoticed rationale for moral action. Of course, there is a risk involved in trusting that the cosmos still has the potential for more-being and wider beauty. The risk, however, is not unreasonable. After all, the universe has already been in the business of intensifying being and beauty for fourteen billion years. You can find immediate evidence for this increase, as I have been suggesting all along, simply by becoming aware of the splendid cosmic outcome you have come to know and value as your own mind.

Anticipation wagers that people will be more naturally inclined to live responsibly if they believe that the universe to which they belong has an indestructible meaning than if they thought of it as inherently pointless and destined for absolute death. If we could understand each transient life—and not just human life—as an unrepeatable participant in a *cosmic* journey toward deeper being and beauty, it could add considerable weight to moral motivation and to the commandment to love. For this reason, anticipation considers an unfinished universe to be a most favorable context for the cultivation of virtue. Here I need to remind readers again of points made earlier: unless the *whole* story of the universe is gathered into the indestructibility and ever-intensifying beauty of rightness itself, it cannot merit our unwavering trust.

The main issue here, therefore, is not so much whether humans will be moral, but whether they will have something worthwhile to live for, some goal or cosmic destiny to fulfill. Religiously speaking, enthusiastic moral action—if it is to be sustained over many generations—requires a guarantee of the lasting significance of our lives and endeavors. The primary function of religion in an unfinished universe, accordingly, is not to draw up lists of right and wrong behavior or to hold out the prospect of rewards and punishments. Rather it is the role of religion to reassure us that the fruits of our labor will not be lost forever. If religion, in addition, gives us specific moral teachings, as at some point it must, it can issue them effectively only after, and not before, it lays out a vision of the universe in which our good works can make a lasting difference.

The archaeonomic vision cannot offer any such promise. Indeed, it formally excludes any reason for trusting that life has a permanent meaning. So, any moral incentives and actions that arise from its reading of the universe can only be futile, at least in the long run. The analogical vision is problematic too for failing to consider how much our moral lives mean if we experience them as contributing to the ongoing story of an unfinished universe. An anticipatory reformation of religious and cosmic sensibilities, by contrast, opens up a meaningful future for human action in the setting of a universe that is still in the making. By understanding our moral lives cosmically, it gives new meaning to our works of love and lays the foundation for an intergenerational pursuit of goodness.[18]

CHAPTER TEN

Wrongness

Surely there is no one on earth so righteous as to do good
without ever sinning.

—*Ecclesiastes* 7:20

RELIGION HAS TO DO not only with rightness but also with wrongness, including the wrongness in religion itself. People of all times
and places have known that the world is not right, or at least not
quite right, and informed religious believers are aware that both they
and religion can go wrong too. The awareness of wrongness, however, cannot occur apart from an intuition of rightness with which to
contrast it. The experience of wrongness would not oppress us so
much unless we already had at least a vague sense of the alternative,
and it is especially in religion, with all its ambiguity, that people for
ages have consciously awakened to the reality of rightness. Even apart
from religion, however, sensitivity to rightness is an inescapable aspect of human consciousness. As rightness dawns, it shines on both
those who welcome it and those who do not, and it does so without
discrimination. We may allow it to enlighten us or we may turn our
backs to it, but in either case its beams shine through. We feel their
warmth, sometimes uncomfortably, even when we deny that rightness is real.

Whenever we aspire to right understanding, right action, and right satisfaction we put on display our fundamental openness to the dawning of rightness, whether we are formally religious or not. Religion, however, has been meaningful to people because it promises a final victory of rightness over wrongness. In biblical religion, for example, God is thought of as the one who will finally "make things right" once and for all. This is, in part, the meaning of "justification" (*sedeqah, dikaiosune*). Following his Jewish ancestors, the Christian apostle Paul used the word "sin" when referring to the powers of darkness that oppose justification, but he taught that in Christ an infinite goodness has in principle set things right for good. Religious myths have always drawn a sharp distinction between the actual world where evil exists and an essential or ideal world where wrongness is overcome. Islamic, Hindu, Buddhist, and other great traditions all distinguish between an unredeemed state of existence and one in which darkness is vanquished by the full rising of rightness.

Analogical religion has been especially appealing because it marks off a realm of timeless rightness sharply from that of wrongness. It offers the prospect of a decisive deliverance from physical suffering, moral evil, disintegration, and death. Archaeonomic naturalists, most of whom espouse materialism and cosmic pessimism, are also aware of wrongness in the world, but they deny—ironically in the name of right understanding—the possibility of an eventual conquering of wrongness by rightness. Aligning themselves with the ancient tragic vision of existence, they profess to be certain of the inescapability of fate and the finality of death. A select few of them find a degree of comfort in the internal swell of courage they experience in facing up to the absurdity of a pointless universe. Cosmic pessimists formally insist, however, that the eventual physical collapse of the cosmos will inevitably extinguish once and for all the wick of life and the light of consciousness. Since everything eventually perishes, wrongness will prevail in the end.

The anticipatory reading of nature that I have been featuring is also aware that something is deeply wrong with our world. Wrongness in this perspective, once again, has something to do with the fact that the universe is still in the process of being born. The universe, at least for now, is not yet perfectly right or fully intelligible because it is not yet fully actualized. Wrongness, then, shows up not because time is a dangerous departure from eternity, as analogy claims, but because the narrative of the universe is far from finished. The universe's incompleteness is a condition that leaves open a space in which wrongness may exist. Acknowledging the fact of an unfinished universe does not mitigate the fact of perishing and the monstrosity of evil, but it does situate the human protest against wrongness, and the universal hope for redemption, in a new intellectual and spiritual context.[1]

In its response to the wrongness that exists inevitably in an unfinished universe, anticipation looks for a fulfillment *of* the temporal universe rather than a refuge *from* it. It cannot ignore the extensive cosmic narrative taking place between the long-ago and the not-yet. At present we find ourselves somewhere—we know not exactly where—between what is past and what is yet to come. Anticipation invites us to turn our hearts and minds toward the horizon of a redemptive rightness that has yet to arrive in its fullness. It feels anguish in the face of evil, especially the injustice and violence wrought by human beings, but it trusts that the present state of the cosmic drama, of which human history is a part, is not final. Things may yet be made right. The world, after all, is still coming into being, and so it cannot be expected realistically to have already reached perfection. As long as the world is still being born, each present moment in the birth process is necessarily alloyed with incompleteness. From the perspective of anticipation, then, wrongness finds a foothold in the shadow side of an unfinished universe. In that same universe, however, it senses an opening to deliverance.[2]

The incompleteness of the universe allows for both natural and moral wrongness. Natural wrongness consists in part of the simple fact that things perish.[3] As I shall observe later in this chapter, natural wrongness resides also in the excess of suffering that accompanies the whole story of life and not just human history. Moral wrongness, on the other hand, is the willful human substitution (or mistaking) of a partial or temporary achievement—whether in our own lives or in the universe—for final coherence. Moral evil, in one way or another, is always an impatient mistaking of a limited good for a full and final rightness. In a world that is still being born, moral evil is the consequence of our confusing a local, present, or past state of things with a wholeness yet to arrive.

Along with the archaeonomic vision, anticipation frankly acknowledges the nascent and developmental quality of the cosmos, but unlike cosmic pessimism it does not assume that present imperfection or wrongness is final—especially since the story is still going on and not all the evidence is in. Anticipation, therefore, lives on the virtues of trust and patience, waiting in silence and sometimes in the shadows, for a deeper narrative coherence to dawn on the horizon of the not-yet. The waiting is not passive. It is an active and opportunistic expectation of the arrival of new possibilities presently unseen. It stays awake waiting for the dawn. Together with the analogical vision, anticipation expects rightness to prevail over wrongness, but it rejects the belief that a fully finished world already exists, or has ever existed, either in the temporal past or in timeless splendor "elsewhere." Moreover, as we noted in the previous chapter, an anticipatory morality fosters the life of virtue by offering us an incentive to contribute, in whatever large or small ways we can, to the ongoing creation of the universe.

To keep the future open, anticipation questions the analogical religious idea of an original perfection or stationary plenitude of being. Such an idea seems to imply that nothing more can be accomplished in time than already exists from eternity, so it can foster both moral and religious passivity. Cosmic history would in that case be inconsequential and the passage of time futile in the long run. From the perspective of anticipation, analogy's assumption that rightness has already been fully realized can be ethically disheartening. By suppressing the sense of a new future for the universe, analogy diminishes the importance of moral action and the zest for living.[4] Evil in that case would only be evaded, not conquered.

From the perspective of anticipation evil does not have its origin in a prehistoric fall or cosmic catastrophe. Wrongness is not the consequence of a mythic rebellion during which a bad world splits off from a good one. Rather, wrongness is a quality that may befall any universe as long as it is still on the way. The wrongness in our world, including the evil attached to religion itself, is intolerable and often unspeakably dreadful, but its existence is logically consistent with the fact that the world, since it is not yet fully realized, is not yet fully intelligible, good, or beautiful. Anticipation, it follows, expects not only the universe but also religion to undergo constant reformation. After all, since religion is part of an unfinished universe, it cannot yet be finished either. It may play host to wrongness whenever it sanctions escapism and intolerance. Religion in an unfinished universe can decay into impatience or quietist indifference, but it can also flourish—at least occasionally—in humble awareness that it too can become more. From an anticipatory perspective, rightness—the fullness of being, meaning, truth, goodness, and beauty—is still rising on the horizon of an unfinished universe. Anticipatory religion, accordingly, looks toward the final vanquishing of evil not in the soul's departure to another world but in the full awakening of both the universe and faith, a drama that at present is far from finalized.

Archaeonomy, by comparison, gives wrongness a lasting place within the total scheme of things. The typical scientific materialist, adhering to a metaphysics of the past, takes wrongness to be finally irremovable because the totality of being is destined by what-has-been to end up in a state of elemental, lifeless disintegration. The academically sanctioned archaeonomic worldview spurns any notion that the universe may be a transformative drama in which light may eventually prevail over darkness. It professes to be highly empirical and realistic, but it leaves out of its survey of nature the fact that the cosmos is still in the process of becoming and that there are possibilities yet to be realized. Even though the scientific materialist may be as personally sensitive to evil and as outraged by suffering and death as any other human being, the archaeonomic worldview raises no effective protest against the prospect of eventual decomposition. It assumes without question that final catastrophe is inscribed indelibly in the entropically slumping cosmos from the outset. Consequently, the archaeonomic vision, including materialist versions of Big History, can offer no metaphysically unassailable reason why we should exert ourselves intergenerationally in struggling against inevitable decay. Wrongness wins out in the end no matter what we do in the meantime.

Analogy, as we have seen, is aware of wrongness too, but its reaction to it can easily decay into a perfectionist, impatient avoidance. Rather than staying with cosmic time and allowing its moments to add up to something unprecedented, analogy seeks an exit from time as soon as the soul has proven its moral worth. Wrongness shows up in religious consciousness, according to the analogical perspective, only because the soul is still tied, perhaps by a dim kind of remembrance, to the timeless perfection from which historical time seems to be a fall. Today this would imply that the entirety of cosmic history is a deviation that needs to be undone rather than perfected. By disciplining the flesh, practicing a spirituality of contemplation, and eventually by dying, the soul finds release from the temporal world and returns to the timeless sphere of initial completeness. Rather than waiting vigilantly for rightness to be concretely actualized in a transformative cosmic/historical drama, analogy ideally wants to extract suffering human subjects as cleanly as possible from their imprisonment in matter. By closely associating wrongness with temporality, it looks not for a fulfillment *of* cosmic time but instead for a one-way exit *from* the transient world.

The sacramental leaning of the analogical perspective no doubt allows that the material world can give us glimpses of an eternal goodness. As I have noted often before, however, analogy is mostly immune to time and cosmic history. It cannot easily make a space in its metaphysics for the new cosmic discovery of the reality and creativity of deep time. This, I believe, is one of the reasons why religious people who are still attached to analogical spirituality fail to pay much attention to contemporary science, especially evolution. To analogy's most devout adherents the narrative journey of matter and life is at best a curiosity and at worst an annoying detour on the human journey to eternity. Analogy is indifferent to the possibility that the story of the universe has a meaning still in the making. Consequently, it subtly tolerates temporal wrongness by allowing the physical universe to fall away into final insignificance. In its most otherworldly extremes, analogy promises the redemption of the human soul only by allowing the temporal, physical universe to be eventually sloughed off unredeemed.

Once nature's sacramental function has been exhausted, according to analogy's typical traditional perspective, the physical universe left behind by the soul's flight to heaven is a mindless and soulless wasteland. Tragically, the desolate cosmic residue left behind by the soul's exodus from time has now become the foundation of modern archaeonomic naturalism. The idea of matter espoused by materialists today is, in considerable measure, a leftover from centuries of otherworldly optimism that had already in effect drained nature of its vitalizing temporal sap. The cosmic pessimism of so many modern intellectuals, it turns out, is a cultural by-product of the implicit despair about the physical universe that had been tolerated for so many centuries by otherworldly religious readings of nature.

Neither analogy nor archaeonomy, therefore, can feel fully the urgency of expectation that wrongness stirs up in organisms that strive and suffer in time. Neither of them appreciates the degree to which the present experience of evil calls for a transformation of the whole cosmos and not just human souls. Anticipation, on the other hand, refuses to separate the experience of wrongness from a universe that is not yet fully real. Anticipation reconciles its adherents not to a final victory of death over life but to a cosmos in which more being, intelligibility, goodness, and beauty may yet emerge dramatically in an always-arriving and not yet fully actualized rightness.

The Wrongness of Evolution

Archaeonomic naturalists, in spite of their formal denial that rightness is real, are not personally insensitive to the wrongness in nature. For them wrongness is real even if rightness is not. Ever since Darwin their aversion to evil comes out especially in passionate protests against the cruelty and indifference of biological evolution. Evolution, they observe, contradicts every sensitive person's ethical sense of rightness. Citing the way in which natural selection wastefully discards organisms and whole species, both Charles Darwin and his chief advocate T. H. Huxley concluded that wrongness is built into the evolutionary process and that nature therefore cannot be trusted to tell us humans how to live rightly. Contemporary evolutionists usually agree.

During his studies at Cambridge and before his acclaimed sea voyage on HMS *Beagle* (1831–36), Darwin had little difficulty accepting the analogical religious belief that the universe reflects a transcendent rightness, however imperfectly. Instances of adaptive design in the natural world pointed sacramentally to an order of eternal goodness. That birds are fitted with wings, that fish can see clearly underwater, or that humans are endowed with limbs and brains to cope with their environments had been a source of religious astonishment for centuries. Pre-Darwinian naturalists could not imagine how the adaptation of an organism to its environment could be anything other than an expression of infinite intelligence and supreme rightness.

The young Darwin had devoured this point of view eagerly, but that was before he learned by close observation that life is a long, twisted, and bloody affair. During the nineteenth century, after geologists had shown that Earth bears the blemishes of a catastrophic past, came the news that the life-process on this same planet is on a hazardous journey of its own. Life is tough, as humans have always known, but the full tragedy of its long struggle had been hidden from view until Darwin came along. Organic design now turns out to be not proof of divine guidance, but the end product of a long, impersonal filtering process during which all nonadaptive instances of life have been discarded ruthlessly. In the drama of life most organisms and species have been left behind to die, unfulfilled and now unremembered. Design, Darwin discovered, is only apparent, not real. Living forms are accidental products of an underlying wrongness. Consider all the maladapted organisms that had to be discarded before a few survivable versions could slip into the story. Look also at the crudeness and engineering inefficiency of natural selection. And, of course, don't forget how much time has been misspent in life's pointless experiments.

Darwin was unusually sensitive to natural selection's inefficiency and its indifference to suffering and waste. He could not get over the fact that many more offspring are typically produced in each generation than ever reach maturity, and that most organisms are mercilessly exterminated before having the opportunity to reproduce. Only a select few survive long enough to bear offspring, he noticed, and they have been granted this opportunity only because they "just happened" to be better adapted to their environments than others. It is hard to imagine anything like intentional benign design behind the process.

Troubled by the excess of struggle and pain in life's evolution, Darwin gradually abandoned his earlier belief that nature is rightly governed. Endowed by birth and training with a refined reactivity to wrongness, he took special offense at the cruel way natural selection works. He was morally disturbed, for example, that the female of a particular species of wasp deposits its eggs inside the bodies of living caterpillars so that newly hatched larvae could nourish themselves on their host while it is still alive. Such an adaptation may be inventive and even ingenious, but it seems perverse from the point of view of human standards of moral rightness.

Contemporary "neo-Darwinian" evolutionists react no less censoriously than Darwin and Huxley to the uncaring way in which evolution works. The late George Williams, one of the luminaries of contemporary biology, referred to nature as a "wicked old witch" for the inhumane manner in which it brings about new species of life. "With what other than condemnation," he asked, "is a person with any moral sense supposed to respond to a system in which the ultimate purpose in life is to be better than your neighbor at getting genes into future generations?"[5] In a similar spirit of moral protest, the paleontologist Stephen Jay Gould declared that the "cold bath" of Darwinism should deliver us once and for all from the illusion that we can learn anything about rightness from looking at the natural world:

> When we thought that factual nature matched our hopes and comforts . . . then we easily fell into the trap of equating actuality with righteousness. But after Darwin . . . we finally become free to detach our search for ethical truth and spiritual meaning from our scientific quest to understand the facts and mechanisms of nature. Darwin . . . liberated us from asking too much of nature, thus leaving us free to comprehend whatever fearful fascination may reside "out there," in full confidence that our quest for decency and meaning cannot be threatened thereby, and can emerge only from our own moral consciousness.[6]

Likewise, Philip Kitcher, a thoughtful philosophical expert on Darwinian evolution, remarks that "a history of life dominated by natural selection is extremely hard to understand in providentialist terms." Evolution, in other words, is simply wrong. "Indeed," Kitcher adds, "if we imagine a human observer presiding over a miniaturized version of the whole show, peering down on his 'creation,' it is extremely hard to equip the face with a kindly expression."[7]

The Cosmic Meaning of Compassion

Most offensive to our compassionate evolutionists are three components of Darwin's account of life: first, the high degree of accident or randomness in the origin of species; second, the unfairness of the "law" of natural selection that allows only a few organisms to survive and reproduce while the majority lose out in "the struggle for existence"; and third, the seemingly wasteful amount of time it has taken for life and mind to make their entrance into the universe. Darwin himself, reflecting on the randomness, mercilessness, and sluggish pace of evolution, could not reconcile evolution with rightness, at least in the context of the highly moralistic Victorian times in which he lived.

Nevertheless, even though his ethical sensitivities were wounded by what he beheld in the story of life, in the end Darwin could still attribute a kind of grandeur to it all. A strange quality of aesthetic rightness became vaguely transparent to him even amid the engineering inelegance and ethical heartlessness of the process. The feeling of awe Darwin felt in reflecting on the "tangled bank" of life seems to reflect his implicit awareness of the dramatic quality of nature. Although Darwin may not have been formally aware of it, his masterpiece *On the Origin of Species* is the most significant contribution any scientist has yet made to the modern discovery of nature's narrativity. We can begin to understand life, in other words, only by telling a story about it. And if the life story is still going on, perhaps there is much more to it than we can make out at present.

Given the warped ways of evolution, nevertheless, how does the story of life fit into our general portrait of a cosmos awakening to the dawning of rightness? In responding to this question I believe we need to focus on the "dawning" as much as on the rightness. Nothing in the history of science, as far as I know, has raised the question of nature's rightness more agonizingly than the discovery that life evolves and new species arise by natural selection. Let it be noted, however, that there is much more to evolution than competition, pain, and waste. The life-story also includes the often ignored elements of creativity and cooperation. The process has never been unambiguously wrong, and it has, after all, given rise to conscious beings who aspire to rightness. Still, at least to many empathetic naturalists the Darwinian account of life has put an end to any reasonable harmonizing of the cosmos with rightness. Above all, it has destroyed the idea that nature is guided by a providential divine "plan" or "design."

What usually goes unnoticed amid all the moral condemnations of evolution is that the compassionate evolutionists themselves are part of a universe still coming into being. Just as their intelligent subjectivity is a blossoming of the cosmos, so also is their attraction to moral rightness. The evolutionists' own outrage against the excessive suffering of living organisms is itself a palpable expression of the universe's awakening to rightness—in this case through the medium of their own moral sensitivity. During nature's seemingly heartless evolutionary journey, the cosmos has without question given rise to a rich diversity of life, and recently to organisms endowed with an ever-increasing capacity to empathize. Evolution has produced increasingly intense forms of subjectivity, and in human beings subjectivity has evolved into an interior aspiration to meaning, truth, goodness, and beauty. This rich vein of subjectivity, as I have been insisting all along, is as much a part of the cosmic journey as anything else science seeks to understand. So, too, are the human cultures that have nurtured a sense of rightness and reduced the pressures of selection. The evolutionists' own attraction to rightness, in other words, is part of the inside story of the universe. We cannot divorce it from the axial awakening to rightness that I have associated with the emergence of religion in cosmic history.

To evolutionary naturalists, my insistence that their own moral passion is part of a cosmic awakening to rightness may seem surprising. This is because most of them—still unconsciously tethered to an analogical vision of nature—do not yet really believe that their own subjectivity is part of the real world. Uncritically embracing the modern materialist myth that subjectivity is not part of nature and that it may not even have any real existence at all, they condemn the "outside" evolutionary story of life for not measuring up to moral standards that have quietly shaped their own interior lives. Scientifically speaking, compassionate naturalists such as Williams, Gould, and Kitcher are compelled to acknowledge the seamless physical and historical continuity between humans and the cosmos. Yet in the very act of condemning nature for sponsoring evolution, they implicitly exempt their own moral and cognitive lives from being a significant part of the cosmic story.

I believe it is their own unconscious and anachronistic attachment to the prescientific cult of analogy that keeps our renowned evolutionists from admitting that their own compassion is part of a cosmic awakening. The irony is glaring. The gentle-souled Darwinians are absolutely certain of the rightness of their moral condemnation of evolution for being uncaring. They fail to tell us, however, on what ground they are standing when they pass judgment on the world's wickedness. On the one hand, they condemn the wrongness of natural selection. On the other hand, they tell us that this same nasty mechanism is the creator of conscious beings and cultures endowed with an infallible sense of moral rightness. Even in the act of claiming that the evolutionary process is aimless and immoral, they take their own moral sensitivity and mental acuity to be above criticism. Only an unconscious, inherently dualistic, attachment to analogical mythology can account for their sense of being so separate from the universe that gave birth to them and the cultures that have shaped their moral lives.

To anticipation, on the other hand, the highly developed human capacity to rage morally against the wickedness of evolution is one more key to what the cosmic story is really all about—the awakening to rightness. Once we fully admit the continuity of human existence and human cultures with the rest of nature, we are authorized to learn about the universe no less by meditating on acts of human kindness and cultures of moral compassion than by looking at the causal series of events that evolutionary biology isolates for investigation. Allowing that rightness dawns only gradually, anticipation does not need to separate the messy Darwinian chapters narratively from the recent human moral awakening to rightness. Natural selection and the human capacity for compassion are both part of one long story of cosmic awakening. While analogy locates the fullness of being in an eternal present, and archaeonomy locates the matrix of all being in a fixed and unfeeling past, anticipation proposes that the universe has not yet fully awakened to rightness because it is still coming into being.

Awaiting Coherence

For anticipation, the cosmos has always been undergoing a dramatic transformation in the course of time, and it may continue to do so in the future as well. The universe is now barely awakening to rightness, but this awakening nonetheless is a key to the universe's identity and meaning. The universe is not a mere reshuffling of mindless material particles, as the archaeonomic naturalist maintains. Nor is the temporal universe simply a dim reflection of a timeless fait accompli, as analogical religion assumes—nor merely a shadow of mathematically perfect principles, as analogical physics takes it to be. Rather, the universe is a dramatic, transformative, temporal awakening to a rightness that is only gently, and never forcefully, disturbing its sleep.

Anticipation therefore fully acknowledges the imperfection of nature, including design flaws and evolutionary suffering. Yet it makes room for redemption in a narrative coherence not yet attained. Unlike its two rivals, anticipation aligns itself with a universe that is just now opening its eyes, so it patiently awaits the fullness of daybreak. This long-suffering vigilance allows anticipation to face fully the wrongness of evolution without having to accept its finality. It protests the wrongness in evolution, but it is aware that idealistic dreams of perfection cannot be realized in an instant. To all of the compassionate evolutionists who contend that the world should have been designed perfectly from the start, anticipation can only answer that a world fully fashioned from all eternity would be a work of magic rather than a significant story able to carry a meaning. If the cosmos had been perfected instantaneously from the start, there would be no temporal interval between origin and end. Hence there could be no narrative and nothing that could give purpose to cosmic process. And there could be no life, no future, and no human freedom, all of which can exist only within the horizon of what is not-yet.

Since Darwin's day, moreover, the story of awakening has become increasingly intriguing as science has linked the long story of life and mind ever more intricately to the even longer 13.8 billion-year-old cosmic process. Since the whole universe is a far from finished project, the question of the rightness of evolution is no longer reducible to whether nature conforms here and now to a narrowly human sense of decency, design, and moral order. The biblical book of Job and the religious wisdom of many traditions instruct us that the universe anticipates a rightness that is incomprehensible in terms of our narrow, culturally conditioned moral sensibilities. "Where were you," Yahweh addresses Job, "when I laid the foundation of the earth? Tell me, if you have understanding" (Job 38:4). In tune with the author of Job, anticipation refuses to judge the universe by the strict tribunal of our own ethical and cultural standards. Its reluctance to condemn seems all the more appropriate if the universe is still in the making.

Both the analogical and archaeonomic readings are too impatient and perfectionistic to make room for either Job or the new cosmic story. Centuries of immersion in the analogical vision of existence have stamped on the minds and souls of all of us, Darwinians along with the friends of Job, a puritanical expectation that nature should be an immediately perfect implementation of rightness. The universe, as we are now discovering, does not work this way. It has a dramatic rather than an architectural constitution, and the drama carries an insideness inaccessible to objectifying comprehension. The universe is not the product of a fixed plan but instead a long story whose intelligibility, unlike that of a design, cannot be laid bare instantaneously. Its meaning can only be awaited with disciplined forbearance. Demanding that the universe appear initially in the form of a perfect design rather than as an unfinished drama is asking for it to be dead on delivery.

It may take ages before religious and philosophical traditions acknowledge that an anticipatory orientation corresponds more closely than archaeonomy or analogy to what we now know about the universe. Scriptural literalists and evolutionary naturalists alike, different though they may be in other respects, will continue to associate evolution with wrongness as long as they measure it in accordance with analogical standards of engineering excellence rather than anticipatory criteria of narrative coherence. Anticipation does not expect to render the wrongness of evolution and the monstrous evils in human history intelligible in terms of any present or past understanding of things. There is no religious or systematic worldview available that can now make complete sense of wrongness, and even if there were, instead of conquering wrongness it would subtly legitimate it. Wrongness cries out not for reason but for redemption. It waits for a rightness not solely of our own making, a rightness whose rising even the still undeveloped eyes of faith can barely make out through the morning mist.

Finally, it need not be forgotten that religion itself has arisen in the context of an unfinished universe, and so it remains as far from journey's end as the universe that bears it. Tragically, in religion a turning away from rightness occurs almost as soon as the first light begins to fall. The rays of rightness are too bright to be received instantaneously, so religious persons and communities have devised ways to dim their all too dazzling gleam. Thus wrongness is no less the shadow side of religion than of the whole awakening universe. A turning away from rightness leads religion to sink back into magic and idolatry, into substituting a single sectarian episode for the whole story. Religion in that case turns into obsession with certitude and doctrinal exclusivism. At the extreme, it sanctions the violent destruction of others upon whom the light does not seem to have fallen. The evil wrought by religious people, often in the name of a shallow perfectionism, is perhaps the most palpable demonstration we have of the fact that the cosmos has not yet been aroused fully from its long sleep.

CHAPTER ELEVEN

Happiness

It fortifies my soul to know
That though I perish, Truth is so;
That howsoe'er I stray and range,
Whate'er I do, Thou dost not change;
I steadier step when I recall
That if I slip, Thou dost not fall!

—ARTHUR HUGH CLOUGH

RELIGION IS ALSO ABOUT happiness. Finding happiness, however, cannot be the first aim of religion. If happiness comes at all, it can only be as a by-product of being grasped by rightness. Happiness is a kind of enjoyment, but not every kind of enjoyment is right. Wrongness is enjoyable too, for otherwise we would not be attracted to it. This is why religion instructs us to distinguish between right and wrong enjoyment. In Buddhism, for example, the deliberate pursuit of enjoyment leads not to true happiness but to dissatisfaction (*dukkah*). To experience right enjoyment (*nirvana*), we need to let go of greedy enjoyments (*tanha*).[1] To be truly happy we must avoid seeking happiness as though it were something we could possess. For its dawning we need to wait in silence.

Biblical religion also warns us that the pursuit of idols—of things that are too small for us—leads to unhappiness. Psalmists and prophets alike turn us away from idolatry toward the true enjoyment of living in the expansive presence of the infinite: "As a deer longs for flowing streams, my soul longs for you, O God. My soul thirsts for God, for the living God" (Psalm 42:1–2). "One thing I asked of the Lord; that will I seek after: to live in the house of the Lord all the days of my life, to behold the beauty of the Lord, and to inquire in his temple" (Psalm 27:4). Hindu Vedanta is also aware of how easily we become entangled in the restricted, shallow world of mere appearances (*maya*) and hence forgo the bliss that comes from rapt communion with ultimate reality (*satchitananda*).

Religion, generally speaking, teaches that we can find true happiness only by allowing ourselves to be carried away by rightness, the ultimate goal of our longing and an ideal not of our own making. Religion is the grateful awakening to rightness, a horizon that we cannot grasp but that can grasp us and free us from narrow associations that bring unhappiness. The important question, then, is not how to find happiness but how to conform to rightness. Happiness will take care of itself.

Nevertheless, happiness remains an explicit goal for most people, even if their efforts to find it usually meet with frustration. If religion had failed to promise happiness, it would never have taken hold among people and endured for so long. Yet the happiness promised by religion has many meanings, and religious traditions lay out different pathways toward it. Furthermore, not every moment in a devoutly religious person's life is filled with happiness. Unhappiness always threatens to break into our lives, and religious people can experience long periods of fear, doubt, and emptiness. "It belongs to the depth of the religious spirit to have felt forsaken, even by God," says Alfred North Whitehead.[2] Being religious, then, is not a guarantee of immediate satisfaction, for it also includes episodes of feeling estranged from rightness. In an unfinished universe how could it be otherwise? Can the thirst for happiness ever be finally satisfied short of the fulfillment of an entire universe?

At its best, religion is an active and patient waiting for rightness, a goal that lies mostly beyond the limits of direct experience. At its worst, religion is an impatient demand to be carried off immediately into a wonderland of gratification out of touch with the dynamics of a still emerging universe. Religious longing easily turns into impatience, an unwillingness to wait attentively and actively. Religion may decay into pietistic escapism or an obsessive quest for shallow titillation, including delight in the slaughter of innocent people who do not fit into one's religious worldview. When this happens, religion has degenerated into something unimaginably evil.

Complicating the question of happiness and religion even further is the fact that countless nonreligious people testify to having found true happiness only apart from religion. The novelist and philosopher Albert Camus, for example, argues that even Sisyphus, the mythic exemplar of all who feel crushed by the universe's indifference, may be happy. Cleansed of hope and fully conscious of his absurd fate—that of repeatedly rolling a rock up a hill only to watch it roll down again—Sisyphus feels superior to the gods who wallow in luxury and have no need of courage. The "absurd hero" experiences happiness in the surge of vitality that wells up in the human spirit when it faces impossible tasks intrepidly.[3]

Happiness, as Camus insightfully implies, is closely associated with feeling fully alive, but for most human beings vitality requires a sense that life has a meaning.[4] Animals may feel fully alive simply by letting their instincts carry them away. Humans, however, feel most alive when they allow themselves freely to be carried away by a cause to which they can dedicate themselves day after day. Sisyphus therefore can be happy only if he too finds a hidden meaning in facing up to the universe's meaninglessness.

But what exactly is this hidden meaning? To put the question in a contemporary cultural context, how can the scientifically educated cosmic pessimist find hidden meaning in a pointless universe? Having studied for many years the works and biographies of scientific thinkers who consider the universe pointless, and having known some of them personally, I have come to believe that what sustains them and gives meaning to their lives is not a sense of superiority to fate but the sense of being carried away, perhaps unknowingly, by something of incontestable rightness.

To be specific, in their commitment to what they take to be right understanding, cosmic pessimists silently place their lives in the service of an indestructible value. By scorning wishful thinking, they allow their lives to be enveloped, and their thoughts and actions to be judged, by a noble value not of their own making: truth. Truth, for them, is an encompassing horizon to which they humbly bow and which they desire to serve, even if it hurts. Scientific thinkers who suspect that the universe is pointless may feel nevertheless that they are dedicating their lives to the pursuit of rightness. Charles Darwin is a good example. Relentless pursuit of truth—that is, right understanding—is not always the case in science since there are instances of cheating and careless disregard for truth in its long history. For courageous scientific thinkers such as Darwin, however, truth means rightness of understanding, and they are willing to make many personal sacrifices, including at times their own reputations, on its behalf. Their dedication is quiet testimony that rightness, even for the cosmic pessimist, is larger than the universe.

Right understanding, then, is an undeniable ideal, and good scientists feel an obligation to distinguish it constantly from seductive flashes of insight that turn out to be wrong. Not every bright idea, after all, is a true idea. Bright ideas are a dime a dozen, but right ideas come to birth only by steadfast, and sometimes painful, dedication to finding what *is* the case. Right understanding emerges only after cutting away, sometimes reluctantly, at what one would *like* to be true.[5] The passionate pursuit of truth, even if it ends up concluding that the universe is pointless, brings its stewards into the comforting embrace of a nameless but indestructible rightness. If any cosmic pessimists happen to be reading what I have just written and still vigorously oppose my perspective, this reaction can occur only because they too are subservient to a rightness that can expose the "wrongness" of my point of view. They too have a tacit personal trust in an indestructible value that lies deeper than either the visible world or themselves.

Even those who consider the cosmos pointless therefore can find a hidden meaning—and a kind of happiness—by surrendering to the value of truth, even if it is cultivated only through the self-limiting perspective of scientific method. Ask any fully convinced cosmic pessimist if he or she values truth unconditionally, and the answer will almost always be affirmative. During interviews Richard Dawkins, for example, is sometimes asked why he pursues his evolutionary atheism with such passion. His answer is usually given in one word:

truth.[6] Dawkins is not alone among scientific materialists in experiencing a kind of meaning and a sense of personal nobility in allowing himself to be carried away by rightness, even if it is disguised in the self-contradictory creed of "scientism," the (unscientific) belief that science is the only reliable road to truth. The pursuit of truth, after all, is nothing other than the quest for rightness of understanding. Contrary to Camus, then, it is not a sense of moral superiority to fate and the gods, but an encounter with the gracious horizon of indestructible rightness, that allows the light of happiness to shine through at least dimly even in the worlds of Sisyphus and scientism. The hidden prize—even in the most agonizing instances of cosmic despair—is a satisfaction that, even though we perish, "truth is so." Commitment to the transcendent, incorruptible value of truth, as the poet says, is enough to fortify the soul.[7]

A scientifically informed cosmic pessimist, not unlike Sisyphus, can be happy. A courageous personal commitment to factual truth, even if it leads to a formal denial of cosmic meaning and to the repudiation of God-consciousness, can be sustained only if it is accompanied by some degree of satisfaction. The contentment experienced here, however, stems not from the thrill of battling against fate or the universe but from a glimpse of rightness, however filtered its light may be in scientific inquiry. Doing science, therefore, especially if it takes up one's whole career, requires more than genius or a high IQ. It demands the courage to distinguish rightness from wrongness, and courage is a cardinal virtue with which not every gifted thinker is automatically graced. In any case, the pursuit of virtues such as courage, honesty, and humility, along with whatever happiness this pursuit brings, is not at all irrelevant to a life in science. Darwin is so fine an example because of his courageous commitment to the pursuit of truth no matter how painful it was for him to break the news to the rest of us about life's cruelty.

We can appreciate Darwin's tacit surrender to rightness even better if we now examine carefully several of the faith assumptions operative in every noble scientific pursuit of truth. In addition to proper training and native intellectual talent, doing science in the right way requires of the scientist an implicit commitment to a set of value-laden beliefs that do not show up within the horizon of scientific inquiry as such but that nonetheless energize the pursuit of scientific truth. These beliefs usually go unnoticed, but they underlie the personal passion required for a life devoted uncompromisingly to right understanding. For example, any self-respecting scientist has to believe (or trust), first, that the universe is intelligible; second, that truth is worth seeking; third, that honesty, humility, and generosity in sharing one's discoveries are unconditionally right; and fourth, that the scientist's own mind has the capacity to grasp intelligibility and distinguish what is true from what is false. I am confident that any truly reflective scientist will agree that these commitments are essential to doing science rightly.

These expressions of trust, however, may, at least to thoughtful inquirers, give rise to a corresponding fourfold set of questions that science, all by itself, cannot answer: Why is the universe intelligible? Why is truth worth seeking? What makes a scientist's commitment to honesty (and other virtues essential to being a good scientist) unconditionally right? And why should a scientist, especially one who embraces cosmic pessimism, trust that his or her mind, in spite of all its limitations, can grasp intelligibility and make true judgments? Science alone cannot meaningfully answer these four questions because, even before embarking on the adventure of discovery, the scientific inquirer must already have taken for granted that the universe is intelligible, that truth is worth seeking, that honesty is unconditionally right, and that it is right to trust one's capacity to distinguish factual truth from mere hypotheses.

One cannot do science well without believing that all four commitments are right, but science itself cannot say why they are. Concern about how to justify the four beliefs essential to doing science arises only in the minds of serious scientific thinkers who are willing to look inside and pay close attention to their own cognitional activity. Not all people who do science, above all cosmic pessimists, are willing to undertake this kind of interior exploration. As we have seen, it is more characteristic of archaeonomic naturalists to suspect or deny that interiority is real, ironically even their own. If any of them are reading this chapter, however, they should notice that at this moment their own attempts to find out where I have gone wrong here could be carried out successfully only if they too already trust that rightness is real, that intelligibility and truth are worth seeking, that it is good to share one's discoveries with others, and that their minds are capable of finding truth. These commitments are sufficient evidence of having been grasped by rightness, but one can acknowledge it only by daring to look up close and inside.

Cosmic Pessimism and the Meaning of Happiness

It is the human mind's inescapable attraction to rightness, then, that accounts for the hidden meaning in meaninglessness and hence for the happiness even of a Sisyphus. Being grasped by rightness can sustain a sincere thinker even in the throes of a commitment to cosmic pessimism. Happiness in that case does not consist of the psychic buzz that comes from rebellion against fate or shaking one's fists at the gods, but of the sense of being grasped by rightness.

Does true happiness, you may still be wondering, have to be grounded in a sense of being grasped by rightness? I believe so. No doubt some contemporary debunkers of religious belief are not concerned about rightness. They are content to feel a kind of enjoyment in what they take to be their intellectual superiority over the benighted masses of people, and they seem content to enjoy their cognitive and literary eminence.[8] The kind of satisfaction that this sense of superiority brings, however, has little to do with genuine happiness or the vitalizing kind of enjoyment that accompanies a sustained striving for right understanding. The fleeting tingle that occurs when the longing for public acclaim has met with momentary satisfaction, as almost anyone will agree upon reflection, is not the same as the contentment that comes from unbroken contact with "what is so."

True happiness arrives quietly and graciously only as a by-product of our being open to what is indestructibly right and real. All the great religious teachers therefore contrast wrongful, celebratory satisfaction with the more durable peace that comes from facing up to the everlastingness of truth that alone can make us free. The contentment that accompanies cosmic pessimism, however, is not identical with happiness as religion understands it. What gives contact with truth its specifically religious quality is gratitude, the experience that rightness is a gift not owed. Rightness, religiously understood, does not come cheaply and hence cannot be taken for granted. It cannot be owned but only anticipated. Its full reception requires not only patient waiting but also thankful appreciation, a sensibility that sometimes takes the form of devotion to an eternally generous and resourceful "Thou." Right enjoyment, at least from the perspective of anticipatory religion, has nothing to do with taking delight in one's intellectual dominance over the masses of uninformed people. Nor is it the intoxicating sense of privilege and disdain that comes from the feeling of being set apart by cognitive, doctrinal, or moral purity from one's fellow humans.

To be sure, there is something solitary about religious experience too, but if religion at times asks us to spend time in the desert, it is a withdrawal that intends a return. In the experience of anticipatory faith the community of one's fellow humans is never far out of the picture. Cosmic pessimism, on the other hand, offers a mostly private contentment and an almost ungrateful sense of superiority over the universe itself. Think, for example, of Bertrand Russell's flowery celebration of the solitary scientific Atlas who has faced the emptiness of the universe head on and of whom the universe is said to be unworthy.[9] Russell's portrait of human existence in a hostile universe formally rules out gratitude in any but the most limited sense. The isolation he emulates is not absolute, of course, since at least in his own case the pessimist takes the trouble to express himself artfully for his readers' enjoyment. Like Camus, however, Russell locks the lid on Pandora's Box lest hope leak out and lead his fellow mortals to a kind of happiness forbidden by cosmic pessimism.

The "happiness" that Russell and Camus attribute respectively to the mythic figures of Atlas and Sisyphus consists of an esoteric heroic lucidity. Far from accepting the ambiguity of the universe as a mark of its still coming into being, the cosmic pessimist impatiently declares the universe to be pointless through and through, ruling out the possibility of any redemptive future. Cosmic pessimism fails to take into account what it may mean that the universe is still coming to birth and that not all of the evidence can possibly be available yet for making a final judgment. Ignoring the fact that the world is, in most respects, not-yet, the cosmic pessimist identifies "what is so" and "what will be" with "what has been." Pessimism thus arbitrarily ignores the wide-open possibility that new being and deeper meaning may arrive up ahead.

The great teachers of religious wisdom, on the other hand, have arrived at their visions only by letting go of conventional assumptions about truth and happiness. They have been willing to learn that rightness is not what they had initially thought it to be. Unlike Atlas and Sisyphus, whose arriving at contentment in an absurd universe is largely a private and self-satisfied victory, most great teachers whose lives have been touched by truth and happiness generously invite their fellow humans to participate with them in a common enjoyment of rightness. A longing to distribute their hard-won happiness widely and without borders has led great religious teachers throughout human history to seek disciples. The stories of their lives include long desert journeys, both external and internal, that prepared them for awakening, but they were afterward eager to share the fruits of their quest with others.

Indeed, the greatest religious teachers have emerged from their personal trials with a palpable sense of rightness and an accompanying happiness that they were bursting to have others also experience. Think of the stories about Moses returning from the desert, of Jesus emerging from the wilderness to proclaim the good news of the coming of God, of Gautama's experience of enlightenment after almost despairing of ever finding it. We can only imagine the arduous internal itinerary that led Laozi eventually to share his awareness that the Dao that can be spoken about is not the real Dao, or that persuaded the ancient prophets and Muhammad to proclaim in the presence of idolaters the absolute oneness and righteousness of God.

An irrepressible longing to share with others the happiness associated with their visions differentiates the religious appreciation of rightness from the more guarded contentment of cosmic pessimists. Facing up to the indifference of the universe can bring the isolated cosmic pessimist a feeling of satisfaction similar to that of Sisyphus, perhaps, but this is not the kind of redemption that burns to be shared lovingly with others for the sake of building community. Most tragic heroes have not been missionaries eager to make disciples. Nietzsche, for example, insisted on having no disciples. Bertrand Russell, whose impassioned prose articulated cosmic pessimism for a whole generation of scientific skeptics, allows that an individual can build a happy life on the mound of "unyielding despair," but this kind of advice has never turned into a major social movement that leavens the lives of more than a handful of adherents.[10] Cosmic despair may have been well and good for Russell and a select few other human beings, but the kind of contentment that accompanies cosmic pessimism holds little appeal to the majority of humans who live in desperate economic and political circumstances.

Anticipation and the Meaning of Happiness

Archaeonomic naturalism, taking for granted the absurdity of the universe, despises the religious search for happiness as a longing to escape "reality." Analogical optimism, assuming that nature is not getting anywhere itself, interprets religion as a journey to a blissful resting place beyond time, space, and materiality. To anticipation, however, the new discovery that the universe itself is a dramatic transformation locates the religious search for indestructible right-ness inside the larger cosmic journey. Genuine religious happiness, in this context, is not the sort of satisfaction that comes from having subdued fate, nor is it the levity of expecting to escape from the physical universe altogether. Rather, it is the adventurous anticipation of being on the way, along with all humanity, life, and the rest of the universe, toward a rightness that, for all we know, has just barely begun to dawn.

The way of anticipation, consistently informed by the discoveries of science, trusts that a full encounter with what is really real can happen only through an unflagging expectation of what is not-yet. Anticipation reads the universe as having always had the prospect of becoming more than it is or has been. It observes that matter, life, mind, and the world's recent religious awakening to rightness are emergent new epochs in an open-ended cosmic process. Life, mind, and religion are not alien to this universe but are born completely *from* it. The deepest happiness humans can experience here and now in an unfinished universe, therefore, is anticipation itself.

The Christian theologian Jürgen Moltmann makes this point from an Abrahamic perspective. In opposition to those who complain that hope robs us of happiness in the here and now by postponing satisfaction to the future, he responds that in a world that is still incomplete the happiness of the present *is* hope.[11] To expect happiness to be fully consummated here and now is, in effect, to deny that the universe has a future. Such impatience has the effect of narrowing unnecessarily the temporal, dramatic expansiveness of the universe. Anticipation, on the other hand, gives meaning to our waiting and adds lightness to our living with others in time. Before we learned that the universe itself has had to wait, our impatience led us to seek release from the natural world, and sometimes we gave the name "religion" to that impatient withdrawal. The fact of an unfinished universe, however, now allows us to discover a new kind of meaning and happiness in our having to wait. We realize that there is no encounter with rightness apart from our undertaking a long journey with other living beings and with a whole universe toward the unexplored realm of the up-ahead.

In this cosmic adventure human anticipation links up with the caravan of all striving and struggling organisms, including life-journeys that may be going on elsewhere in the universe—or multiverse. Archaeonomy and analogy, on the other hand, are not ready to join the column of hopeful pilgrims. For them the *has-been* (in the case of archaeonomy) and the *is-now* (in the case of analogy) officially leave no room for the enjoyment of journeying with other living and conscious beings toward the newness of what is *yet to come* in the cosmic story. Intellectually as well as spiritually, neither archaeonomy nor analogy can share the expectant happiness of waiting for surprising things, including human redemption, to happen up ahead in complete continuity with the whole cosmic story.

For several centuries now the world of thought all over the planet has subscribed blindly to an archaeonomic vision that absolutizes the fixed past and, because of its implied determinism, denies that anything truly new can arise on the horizon of what-is-to-come. Sometimes respected thinkers have even felt compelled by science to deny the real existence of their own freedom and responsibility. Archaeonomic naturalism, by disallowing anything really new to happen, cultivates a literalist obsession with certitude that rivals the most spirit-suppressing instances of religious fundamentalism. To the scientific materialist, life and mind—not to mention religion—can only be states of deadness "masquerading" in strange and ultimately insignificant disguises that are destined to lapse back eventually into the common grave of all being.

We live at a time in intellectual history, it is true, when physics itself has begun to vanquish materialist and deterministic concepts of nature, when the notion of emergence is struggling to replace mechanism, when the analytical illusion is giving way to a more ecological understanding of the cosmos, and when time is beginning to be taken, once again, as real. Yet the habitually archaeonomic leaning of modern thought remains mostly undisturbed. It has taken up residence in universities all over the globe. Even though nature itself is struggling to burst out of its enchainment to a simplistic physicalism, academically sponsored intellectual life remains as submerged as ever in the deeply ironic project of trying to make the world intelligible by reducing it to the elemental de-coherence of the cosmic past. Contemporary thought, in other words, still looks mostly to what-has-been in search of a safe way to understand the present and future. It has not yet heard the anticipatory summons to a more venturesome reading of the cosmic story, even though strains of anticipation have been intertwined with human culture and religious traditions for centuries.

Those who follow the way of archaeonomy, with its implied ontology of death, seem content to breathe in the aroma of "sweet decay."[12] They may loudly exclaim that they are perfectly satisfied with a meaningless universe, but by consigning the whole of being to an absolute death they arbitrarily suppress the happiness of anticipation that, if they cared to notice, hiddenly animates even their own quest for right understanding. Meanwhile, the numerous devotees of analogy remain satisfied with the prospect of a decisive release at the moment of death from the burden of bodily and cosmic existence. For them otherworldly optimism is happiness enough for now.[13] Generations of religious people have been content to embrace this acosmic ideal of personal destiny, longing to be set free from a life of earthly suffering. I have no wish to disparage the sense of relief that this kind of eschatology holds out to them, but theirs is not the same as the happiness that anticipation offers by situating our own personal deliverance within the larger scheme of a cosmic redemption.[14] We can find a fuller joy, as far as anticipation is concerned, by allowing ourselves to be carried away along with the universe by a rightness to which the whole of finite being is still awakening.

Summary

The happiness of anticipation follows from gratefully allowing ourselves, and the universe along with us, to be drawn into the heart of rightness. An anticipatory faith not only instructs us that this goal is one for which we must wait, but it also implies that "we are stronger when we wait than when we possess."[15] Academically enthroned archaeonomic interpretations of the universe, on the other hand, are not willing to wait. Even though the cosmic drama is still going on—and, for all we know, may at present be very early in its long run through time—contemporary cosmic pessimists are not reluctant to issue a final verdict here and now that the whole process is pointless. They declare, with absolute confidence in their own capacity for truth-telling, that religious intuitions of meaning and redemption are nothing more than imaginative fiction, and that religious faith and hope can tell us nothing about what the universe is really all about.

In an unfinished universe, however, such a judgment is both premature and empirically unsupportable. To restate what I said early on, such a judgment can occur only because cosmic pessimists tend to look only at the outside story of the universe. They remain mostly unaware, or at least suspicious, of the anticipatory striving of living subjects and of the whole universe that bears us. They treat interiority as somehow unreal, even though, ironically, only a deeply subjective anticipation of truth, as well as a supreme confidence in their own cognitive faculties, can sustain them in their own intellectual efforts.

The anticipatory vision I have been recommending here takes our intelligent subjectivity, with its limitless longing for understanding and truth, to be a blossoming of the entire cosmos, not an arbitrary development that adds nothing of substance to the story of the universe. An anticipatory vision thinks of religion not simply as a set of psychological projections or as escapist poetry emanating from conscious beings who cannot put up with a meaningless universe. Rather, our interior life, including our religious aspirations, is a new emergent stage in a cosmic drama. Our own longing for rightness and the contentment that goes along with it are as much a part of the universe as floods and fires. It is through our own subjectivity—mental, moral, aesthetic, and religious—that the universe now carries on its long anticipatory adventure toward fuller being. Cosmically considered, therefore, our religious anticipation of indestructible rightness—that is, of inexhaustible meaning, goodness, truth, and beauty—along with the special kind of happiness that accompanies this quest, is not escapist human fiction but the very pathway the universe is taking toward communion with its ultimate ground and destiny.

CHAPTER TWELVE

Prayerfulness

We love [God], not only with our whole body, our whole
heart, and our whole soul, but with our whole universe.

—PIERRE TEILHARD DE CHARDIN

NOTHING IS MORE FUNDAMENTAL to religion than prayer. In prayer,
according to the religious scholar Friedrich Heiler, "we have revealed
to us the essential element of all religion." Martin Luther declared
that religion is "prayer and nothing but prayer." The nineteenth-
century German theologian Friedrich Schleiermacher wrote that
"to be religious and to pray—that is really one and the same thing."
And Schleiermacher's contemporary, the German poet Novalis,
called prayer "religion in the making." Richard Rothe, a famous
evangelist, added that "the religious impulse is essentially the impulse to pray" and that nonpraying people are "religiously dead."[1]
To his disciples Jesus said to "pray always without becoming weary"
(Luke 18:1).

It is especially in prayer by members of our species, we may now
say, that the universe undergoes its keenest awakening. It seems
likely that members of our species have been praying in one way or
another for the past 200,000 years, and prayer is the main way in
which religion comes to expression for most people even today. Good

works and moral imperatives are part of religion, too, but they are not as definitional as prayer. Religious scriptures and shrines, as we are coming to realize increasingly in a secular age, are hollow relics unless animated by prayer. Religious symbols and stories mean little until they are incorporated into the prayer life and common worship of real people. Even postaxial God-consciousness, notable for emphasizing the radically transcendent and suprapersonal character of rightness, is empty unless it contributes somehow to the life of prayer.

Like every other human trait, prayer merits multidisciplinary study, but my concern here is to understand what it means in the big historical context of an unfinished universe. It may be of interest to undertake biological, psychological, sociological, neurological, and other ways of examining what goes on when humans are engaged in prayer and worship. What I have proposed in this study, however, is that understanding religion—and now prayer—requires also a wider, deeper, and longer perspective, one that locates our spiritual aspirations, including prayerfulness, within the setting of the cosmic story given to us by the past two centuries of scientific discovery. This fresh point of view portrays the arrival of religion as a dramatic new epoch in a cosmic transformation. Narratively considered, as I have been emphasizing, the emergence of religion—and especially God-consciousness—is no less startling than the breakthroughs that occurred in cosmic history when life and mind made their surprising appearances. In the arrival of religion the universe takes on a new identity and self-consolidation by virtue of its awakening consciously to the dawning of rightness. None of the usual scientific interpretations of religion—each of which may have important insights to contribute to our general understanding—has the breadth of vision to capture the drama of what is going on inside the universe during that awakening.

Religion is a grateful response by the universe to the dawning of rightness, and this gratitude overflows in prayers of heartfelt thanksgiving for the world and all its wonders. Prayer, however, is not just gratitude but also petition. When his disciples ask Jesus to teach them how to pray, he instructs them simply to ask, and to do so in the same way a child does: "Our Father . . . *give* us . . ." "Ask," he says, "and you shall receive." While visiting monasteries and other holy places all over the world, we observe devotees from near and far still prostrating themselves humbly in prayer, begging for gifts of many kinds. Asking for help from beyond is as human an act as any others that anthropologists study.

The Cosmic Meaning of Prayer

The Psalms of the Hebrew Bible, the religious practices of Christians, Muslims, Hindus, Buddhists, and the spirituality of native peoples everywhere spill over with petition along with thanksgiving. Discovering what religion really is in the final analysis, then, comes down to probing the meaning and value of prayer. But isn't prayer merely a conversation with ourselves? In the age of science is it reasonable for people to pray as though there were personal powers beyond immediate experience that are able and willing to respond to their requests? And if there is anything to prayer, why are so few of its petitions answered? Why are there no apparent ruptures of the natural course of events that might provide "evidence" that prayer is effective and rightness responsive? Furthermore, isn't the petitionary quality of prayer a clear indication that our species has not yet grown up? In short, then, can it any longer be right to pray? Given our new understanding of the cosmos as an unfinished story, what is really going on when human beings are praying? Let us look one more time at our three, by now familiar, ways of addressing such questions. Since religion becomes most fully concretized in acts of praying, this final chapter can serve as a recapitulation and concentration of the book's main argument.

Archaeonomy. Archaeonomic naturalism in principle rules out from the start the possibility that anything truly new can ever happen, thus rendering the prayer of petition pointless. Archaeonomy's signature feature, we have seen, is not its devotion to the analytical method of investigation but its denial that the cosmos can ever become more than what it has always been. Archaeonomic naturalists assume that in principle the universe is reducible to its primordial subatomic past, so for its many subscribers references to prayer are nearly taboo today. If materialism is true and nothing really new can ever happen, after all, what would be the point of praying? Since the universe had already reached the apogee of its being, ontologically speaking, in its earliest physical state, nothing that goes on in time since the beginning can add anything real to what already is.

Strict archaeonomy leaves no room for prayerful gratitude either. Giving thanks is ruled out since the universe and everything in it, far from being a free gift, either had to exist or just happened to exist. In either case the cosmos cannot have sprung from any gracious matrix of being beyond itself. Whether the universe exists by necessity or chance, or by some strange combination of the two, archaeonomy cannot interpret it as a gift emanating from an infinite generosity. There may be space in such a universe for feeling lucky or fortunate, but not for feeling prayerfully grateful.

This is not to deny, however, that the average archaeonomic naturalist can be moved by the sentiment of gratitude. In actual fact, by experiencing the shock of existing, people in general, including self-avowed materialists, are spontaneously grateful, and most of them in fact expect to be surprised by unpredictable things that will happen to them in the future. Most human beings are tacitly aware that simply existing at all is an unmerited gift. They unconsciously grasp that even though they are responsible for many things they are not the authors of their own existence. So just as they can live meaningful personal and social lives in spite of their formal denial that the universe has a purpose, archaeonomic naturalists can be implicitly grateful in spite of their explicit denials that the universe or their own existence is a gift.

My objective, once again, is not to deny that archaeonomic naturalists can be grateful. Rather, my concern is whether their materialist worldview is logically compatible with the de facto gratitude they and others have for being alive at all. I have argued previously that the cognitive trust that materialists place in their own minds is out of step logically with their archaeonomic worldview, a way of thinking that in principle reduces all minds to primordial mindlessness. In pointing to this self-contradiction, I have not denied that materialists rightly trust their minds. I have only insisted that their materialist worldview cannot provide a logically satisfying justification of that trust. Here also I want to express my suspicion that when scientific materialists find themselves nurturing the sentiment of gratitude, as they often do, this can only be in spite of, rather than because of, their purely materialist interpretation of the world.

The essence of the archaeonomic vision, after all, is its claim that nothing that ever happens can ever be anything more than fleeting new combinations of elemental original stuff.[2] A purely materialist worldview allows for no real future, no freedom, or for anything of lasting importance ever to happen. As to how the cosmos came to be in the first place, the typical archaeonomic assumption is that either it exists by necessity and has been around forever or it "just happened" without any intelligible reason or purpose. In either case its existence is not something for which prayerful gratitude would be called for, or in which prayerful petition would make any sense.

In this respect, the archaeonomic vision amounts to a thoroughgoing project of cutting the prayerful heart out of an awakening universe. Intellectually grounded in that vision, modern secularist thought has covered over every conceivable opening for prayerfulness, and it has done so more self-consciously than any previous experiment in human thought and culture. As I have been saying, however, few archaeonomic naturalists are consistent enough in their sentiments and thoughts to base their personal lives entirely on an explicitly materialist and pessimistic worldview. Neither sunny nor sober cosmic pessimists are ever so completely in tune with the logical structure of their professed worldview that they really consider their own existence completely meaningless or rule out the coming of more-being into their lives.

Nevertheless, one cannot help wonder what powerful attraction resides in the archaeonomic worldview that it would gain so many devout disciples over the past several centuries, a trend that continues today. What is the secret appeal of archaeonomic naturalism and its implied materialist metaphysics that, in spite of its logically self-subversive internal anatomy, so many highly intelligent people are drawn to it anyway? Is the appeal of materialism simply that it formally rids the universe of any room for prayerful gratitude and petition? Or is there something else that makes it so compelling?

As far as petitionary prayer is concerned, one might answer that science and technology have now made it unnecessary to pray for things that we can give ourselves. The human species has now come of age, it is sometimes said, so we may now rely on our own ingenuity and technical expertise. We need not any longer sink to the low level of religious groveling to deal with our deepest needs and desires. And as far as the prayer of gratitude is concerned, it makes no sense to indulge in it any longer since we have only ourselves to thank for the great achievements that allow us to survive and adapt to an uncaring world.

There may be something worth considering in this kind of response, but its general naïveté becomes obvious once we acknowledge that science and secular culture have done nothing to deliver us from the fundamental existential anxieties of fate, death, guilt, and—above all—meaninglessness.[3] Indeed, for most educated people science often exacerbates their metaphysical pessimism. Science now seems to provide unprecedented intellectual support for their impression that the universe is pointless and that any hope for permanent meaning to their lives is destined for disappointment. The contemporary intellectual presumption of cosmic pessimism, therefore, makes it all the more appropriate today that we ask why archaeonomic naturalists are nonetheless drawn so irresistibly to the idea of an impersonal universe, and hence to a vision of reality in terms of which the whole idea of prayer seems offensive.

The charm of archaeonomic naturalism lies first in the ease with which it allows us intellectually to identify everything in the universe with an underlying homogeneity. Atomic physics, for example, posits a "fundamental equivalence" of one thing energetically with another while contemporary biology collapses all living species into a single, continuous river of genes and proteins. It is easier for our minds to conjure up a universe in which everything arose from, and is destined to fall back into, an all-embracing impersonal physical or atomic uniformity than to accept a universe whose very essence is that of struggling to become increasingly new and differentiated during its passage in time. Materialism delivers the human mind of the need to exercise sustained effort in respecting and accounting for qualitative differences among things. Intellectually, in short, it is easier to deny that more-being has occurred at all than to come up with a worldview that takes novelty and differences to be real.[4]

Second, allowing room for more-being would mean that we would have to take time seriously and along with it the possibility that the universe has an unpredictable future for which we can only wait. Were prayer to have any legitimate standing at all, it would imply that the fullness of being could not have been resident *en arche*, that is, in the beginning. Prayer would mean that we would have to wait patiently for a newness of being along with a deeper intelligibility to which we do not yet have access. Prayer would entail giving primacy to the not-yet over the has-been and the is-now. It would imply that the logically appropriate worldview would be one that can be upheld more by the discipline of restrained anticipation than by archaeonomic fixation on the cosmic past.

Third, the appeal of materialist naturalism consists also of the wall of safety it erects against the anxiety of having to deal with subjectivity, an area of being into which archaeonomy is by definition forbidden to go. The allure of an impersonal universe consists partly of its expulsion of insideness from the cosmos altogether. Materialism, we have seen, denies from the outset that the universe has an interior sphere of being, and this denial gives archaeonomic naturalists—and most authors of Big History—confidence that scientific method is in principle capable of uncovering and mastering the totality of natural and human events. In theistic cultures it also legitimates their rejection of divine personality. Since prayer is one of the most inward and intimately personal of all sets of events going on in the universe, making a place for it inside the cosmic story would in effect restore to the natural world the dimension of subjectivity whose theoretical expulsion has been indispensable to the shaping of the modern intellectual world.

Fourth, underlying the cult of an impersonal universe and the explicitly secularist exclusion of prayer lies an almost mystical enchantment with the prospect of the human mind's being dissolved into a final state of mindlessness.[5] The allure of archaeonomic naturalism is not unlike the irresistible pull that pantheism has had on humans so often during our brief history on Earth. Pantheism, formally speaking, is the belief that nature *is* God and hence that nothing can exist beyond nature. Transcendence is an illusion, in other words. By denying that anything exists beyond nature, pantheism is prone to sanction a kind of spiritual passivity that renders both prayer and hope not worth the effort. In closing off the whole world from the horizon of more-being, pantheism in principle renders futile any human struggle toward what is not-yet. In its modern incarnation as materialist naturalism, pantheism assumes that the natural world available to objective, scientific comprehension is all there is and that the universe's original state of physical dispersal and impersonality is the ground and destiny of all being. By confusing metaphysical primacy with temporal originality, archaeonomy in effect identifies the fullness of being with the state of mindless simplicity that existed first in the flow of cosmic time. Translated, this metaphysical confusion amounts to claiming that mind, which arrives late in cosmic process, is essentially mindlessness. Needless to say, this reduction is hardly promising as a way to justify the cognitive confidence on display in contemporary learned apologies for materialism.

Fifth, for many of its disciples much of the appeal of archaeonomic materialism lies in its denial of the scientifically intractable notion of freedom. If the universe is ultimately impersonal and driven by forces originating solely in the past, there is no space in it for human freedom and responsibility. Nor is there room for a genuine awakening to rightness, or any good reason to accept the invitation to personal transformation. Hence there is no justifiable place for a sense of guilt or for asking prayerfully for forgiveness for turning away from rightness. Since, in the archaeonomic reading, nothing in nature is really anything more than the result of physical causes arising from a deterministic past, the materialist feels justified in assuming that freedom is an illusion irreconcilable with the scientific worldview.

Albert Einstein is only one of many scientific thinkers who have believed that affirming the existence of freedom would spell the end of science. To allow a place for either freedom or prayer to a personal deity would mean that the closed, law-driven continuum of cause-and-effect relationships that makes up the world of scientific exploration could in principle be interrupted.[6] Such an interruption would invalidate the whole project and structure of modern science. Science, according to Einstein, cannot work unless everything in the present and future is in principle completely predictable on the basis of what has happened in the past. Since the universe exists by necessity rather than by an underlying generosity, moreover, there is no place for the religious idea of a responsive God—and hence for prayerful petition and gratitude. Since meaningful prayer would entail the reality of subjectivity and freedom, both of humans and God, it can have no place in a scientific understanding of the world. Consequently, to save science the world must be liberated from any association with freedom and the idea of a personal God. No longer can educated people appreciate the universe as a gift whose existence is the result of pure grace and whose real meaning they would have to wait for in a spirit of prayerful patience.[7]

Analogy. Analogical interpretations of the universe have for centuries provided the main religious and intellectual framework for interpreting the meaning of prayer. To most people of faith analogy still makes good sense of both gratitude and petition. By reading things in nature as imperfect representations of an eternal divine fullness distinct from the world, analogy's sacramental vision invites us to appreciate the transient beauty of finite beings both as a gift that can awaken the sentiment of gratitude, and as a window onto a transcendent beauty that invites us to keep asking our whole lives long for infinitely more of it.

Since worldly things are finite symbols of an infinite beauty they are to be enjoyed without being worshiped. The meaning of gratitude, analogically speaking, is that it allows us to receive the finite treasures of nature and human history as gifts rather than as acquisitions that could ever satisfy us fully. Sacramentalism implies that every finite good participates in a timeless and infinite rightness. Analogical spirituality thereby protects everything worldly from being interpreted as an end in itself or as the product of impersonal chance and necessity. Finite things, including the whole natural world, are to be enjoyed as limited symbols of a limitless beauty, not adored as though they were all that is. If we end up adoring them, sooner or later we will stop enjoying them too, because we will have become their slaves. Prayerful gratitude is liberating, then, because it allows us to appreciate things and persons without confusing them idolatrously with the infinite generosity to which they point.

The universe, as read from an analogical perspective, depends on and participates in an eternal present. The meaning of petitionary prayer, then, is that it opens our hearts to the infinite plenitude of being and beauty whose perfection a temporal, fallen world can never adequately represent. To the distress of our souls, greediness—*tanha, shirk,* idolatry—turns us into addicts, robbing us of the freedom that comes from allowing ourselves to be grasped by what is eternal. Attaching our hearts and souls to finite beings may eventually bring a sense of desolation to human subjects whose deepest longing, as Augustine of Hippo reminds us, is for nothing less than the infinite itself. Petitionary prayer at heart, therefore, is not an infantile request for frivolous things to possess, but a large-souled longing for communion with what lies altogether beyond time and space.

Anticipation. No doubt, analogy has been the dominant way of interpreting both the prayer of thanksgiving and the prayer of petition during the prescientific period of human history. From the point of view of anticipation, however, analogy cannot bring to full expression the poignancy of prayer in the context of our new scientific awareness of an unfinished universe. Analogy's crisp distinction between the finite and the infinite is relevant religiously, but to anticipation the world of time and materiality is not a prison from which we need to seek release. Rather, it is a long pilgrimage whose undertaking gives new meaning to every passing moment in its temporal passage. Analogy, by failing to take into account the itinerant quality of nature and the narrative significance of long stretches of time, narrows down the vast, cosmic panorama of things for which we may now be thankful. Additionally, by ignoring the unfinished cosmic story and the painful uncertainty of its still-coming-to-birth, analogical/sacramental prayer cannot feel fully the cosmic and evolutionary drama of struggle and suffering that now adds to human prayer a dimension of pathos previously unnoticed.

There is room neither in analogy nor in archaeonomy for the dawning in the universe of more-being on the horizon of the not-yet. Both analogy and archaeonomy identify the fullness of being with what each takes to be "archaic," but they are locked in mortal combat over the precise meaning of *arche*, a Greek prefix signifying "beginning" or "origin." To archaeonomic materialism, *arche* refers to what has primacy in the order of time, whereas to analogy—or what we may also call the "archetypal" worldview—*arche* refers to what has primacy in the order of being and lies outside of time. Anticipation, on the other hand, in its search for the fullness of being and intelligibility, leans patiently toward the horizon of what is yet to come in the course of time rather than toward what is taken to be "archaic" in either of the other two readings of the universe. It is willing to wait in a state of prayerful expectation for the dawning of more-being.

For anticipation, the universe rests neither on the past nor on an eternal now but on what is to come.[8] Its metaphysics of the future therefore opens up a space for the prayerful expression of both gratitude and petition. The anticipatory prayer of gratitude arises not only from the fact that the universe exists at all but also from excitement that the universe has been set free to become continually more. Anticipation gives thanks that the cosmos is in the present undergoing an intensification of its being and that in the future the world may become more than what it has been and more than what it is now. In an anticipatory setting our prayer greets the dawning of rightness "not only with our whole body, our whole heart, and our whole soul, but with our whole universe."[9]

In each human prayer of petition, correspondingly, a whole cosmic drama is straining toward final deliverance. At heart, cosmically speaking, the prayer of petition *is* the now personalized universe seeking deliverance from the prospect of falling back into its original state of impersonal dispersal. The archaeonomic vision may be content to consign the universe to final annihilation, and the analogical vision may be satisfied to save human persons from their immersion in matter. Anticipation, however, interprets our emergent universe as both an immense dramatic awakening and an equally expansive waiting for an unprecedented future. To archaeonomy the astrophysical picture of an original elemental sea of atomic particulars is the key to understanding where everything will end up. To analogy the plurality of material things is to be exchanged abruptly for an eternal unity. To anticipation, however, the scattered elemental seeds of cosmic beginnings still have the promise of eventually converging over the course of time into more intense unity, more life, more consciousness, and more beauty. No matter how private and provincial the actual prayers of people may be, an anticipatory reading sees in them a whole universe opening itself, always imperfectly, to more-being. Even though archaeonomic naturalists are embarrassed at the fact that most of their fellow humans still pray, anticipation finds in the posture of human prayer—even in all of its childish and petty supplications—a whole universe bending toward the horizon of renewal.

Concluding Reflections

In the awakening of human intelligence, moral aspiration, aesthetic sensitivity, and religious wonder, the blinding sun of rightness is still rising on the horizon of a newly awakened universe. Sensitive poets, philosophers, and religious thinkers of all ages have witnessed to it in many ways, though always imperfectly. Archaeonomic naturalists, having dug their way back into the remote cosmic past, turn their eyes away from the light that graciously illuminates the cosmos they are seeking to disinter. The light of rightness is unobtrusive, but its presence is felt obliquely even in the most impassioned denials of it. So hidden is it that most of the time we fail even to notice it. Yet its light shines on the just and unjust alike.

Sometimes archaeonomic naturalists vehemently deny that this light really exists, although in doing so they assume ironically that it is right to deny it. An identifying trait of religion, however, is that instead of taking this dawning for granted, it glorifies rightness, prayerfully gives thanks for it, and even worships it. More often than not, in its devotion to rightness religion symbolically endows it with qualities analogous to those of human personality, thus making prayerful conversation with rightness an important part of religion. Religion gives rightness a face, if you will, referring to it as Father, Mother, Friend, Companion, Christ, or Krishna. Such personal images convey the sense that rightness is forever faithful.

For what, though, should we pray, and can we honestly expect our prayers to be answered? Before science discovered that nature is narrative, people of the analogical persuasion prayed more for deliverance *from* the universe than they did for the salvation *of* the universe. Now prayer means bringing one's own expectations into alignment with a cosmos that has always been straining to become more. In this new situation, then, in what sense can we expect our prayers to be answered, if at all? Following the line of thought I have been developing in the preceding chapters, I can only repeat a familiar refrain: We can only wait patiently and attentively, without forcing our will on the flow of time. Prayer, as seen from an antici-

patory cosmic perspective, is a request not for magic but for fulfillment, and we do not live in the kind of universe that allows for that
outcome to take place all at once. Hope for the fulfillment of the
universe, moreover, requires human effort and constant struggle
along with prayerfulness.

Archaeonomically, every prayerful request seems unreasonable,
but let us recall why this impression has become so prevalent. The
modern fixation on the metaphor of law, as illustrated by Einstein,
is responsible, at least in great measure, for the naturalistic negation
of prayer. Taking the notion of law literally, archaeonomic naturalists have assumed that the final destiny of the universe is already
settled and that the whole of being rests entirely on the cosmic past.
An anticipatory reading of nature, however, looks at the regularities
of nature as grammatical rules rather than as implacable, dictatorial
forces. In shifting our main metaphor for the universe from mechanism to that of drama, we allow for a corresponding shift from "law"
to "grammatical rules" in our understanding of how the invariant
habits of nature work. By replacing the juridical image of nature's
inviolable laws with that of grammatical rules we can fully embrace
modern scientific discoveries while leaving ample room for the coming of novelty, surprise, more-being, deeper meaning, and human
freedom into the cosmic narrative.

The universe, in that case, is not fixed or frozen by unbending
rules. It is an unfinished story whose meaning is far from having
been set in stone from the start. In the narrative indeterminacy that
anticipation finds in nature there is bountiful unsettled territory for
the efficacy, at least in principle, of petitionary prayer. In asking
humbly for gifts to nourish our lives we are tacitly begging for the
fulfillment of a whole universe. We may discover that our individual
petitions are not granted all at once or in the way we would prefer. Yet
ample room remains for trust that something even more momentous
than we can individually anticipate may still be coming. In this understanding, prayer is not equivalent to asking that the laws of the universe "be annulled in behalf of a single petitioner confessedly unworthy," as Ambrose Bierce sarcastically defines religious entreaty.[10]
Rather, prayer is asking that the unfinished universe—by slowly
weaving an indeterminate sequence of events onto the loom of reliable grammatical routines—may be drawn graciously toward an in-

destructible beauty whose depth and breadth are still undetermined. In a universe now coming to birth, we may stretch out our minds, hands, and hearts in confidence not only that prayers may be answered, but that, if we wait patiently and watchfully, they may be answered in ways far surpassing what we think we need.

Notes

Introduction

1. A significant exception is Brian Swimme and Mary Evelyn Tucker's *Journey of the Universe* (New Haven: Yale University Press, 2011). I strongly recommend this succinct, readable, and scientifically solid work as complementary to the "inside" look into cosmic history presented here.

2. One critic of materialist cosmology, Thomas Berry, whose work is partly indebted to writings of the Jesuit geologist and paleontologist Pierre Teilhard de Chardin, has also interpreted the universe as a story of emerging subjectivity (along with "diversity" and "communion"). See especially his *The Dream of the Earth* (San Francisco: Sierra Club Books, 1988); see also Brian Swimme and Thomas Berry, *The Universe Story* (San Francisco: HarperSanFrancisco, 1992).

3. John Bowker, *Is Anybody Out There?* (Westminster, Md.: Christian Classics, 1988); and John Bowker, *The Sense of God* (Oxford: Clarendon, 1973). I have developed Bowker's perspective at book length in *What Is Religion? An Introduction* (New York: Paulist, 1990).

4. Paul Tillich, *The Courage to Be* (New Haven: Yale University Press, 1952); and Paul Tillich, *Dynamics of Faith* (New York: Harper Torchbooks, 1958).

5. John Bowker, *Problems of Suffering in Religions of the World* (Cambridge: Cambridge University Press, 1970), 227.

6. This is a point made by Pierre Teilhard de Chardin more than by any other scientist or religious thinker, as we shall see. See especially his *The Human Phenomenon*, trans. Sarah Appleton-Weber (Portland, Ore.: Sussex Academic, 1999).

7. Wilfred Cantwell Smith claims that religion exists concretely only where an individual person's life intersects with a historical tradition; *The Meaning and End of Religion* (New York: Macmillan, 1963).

Chapter One. Dawning

Epigraphs: Gerard Manley Hopkins, "God's Grandeur"; all biblical quotations are from the New Revised Standard Version.

1. John Bowker, *The Sense of God* (Oxford: Clarendon, 1973).

2. Bernard Lonergan, *Insight: A Study of Human Understanding*, 3rd. ed. (New York: Philosophical Library, 1970), 10.

3. Karl Jaspers, *The Way to Wisdom: An Introduction to Philosophy* (New Haven: Yale University Press, 1951), 98. I suggest, in keeping with our cosmic approach, that we extend the length of the axial period of religious ferment a bit both before and after Jaspers's 800–300 BCE.

4. I am not excluding the possibility that this cosmic turn happening on Earth has parallels elsewhere in the universe.

5. The Greek term *dikaiosis* is used in the New Testament to specify the rectifying—the "setting right"—that early Christians associated with the work of Christ. See Fleming Rutledge, *The Crucifixion: Understanding the Death of Jesus Christ* (Grand Rapids, Mich.: Eerdmans, 2015), Kindle Edition, loc. 3522.

6. On the question of whether religion has lived up to its own idealizing of unity, see, for example, my *God and the New Atheism: A Critical Response to Dawkins, Harris, and Hitchens* (Louisville: Westminster/John Knox Press, 2008).

7. A good introduction to the Perennial Philosophy is Martin Lings, ed., *The Underlying Religion: An Introduction to the Perennial Philosophy* (Bloomington, Ind.: World Wisdom, 2007). Examples of the perennialist approach include works by Fritjof Schuon, René Guenon, Seyyed Hossein Nasr, and Huston Smith. See especially Smith's *Forgotten Truth: The Common Vision of the World's Religions* (San Francisco: HarperOne, 1992), and Nasr's *Knowledge and the Sacred* (Albany: SUNY Press, 1989).

8. See, for example, Robert Hinde, *Why Gods Persist: A Scientific Approach to Religions* (New York: Routledge, 1999); Walter Burkert, *Creation of the Sacred: Tracks of Biology in Early Religions* (Cambridge: Harvard University Press, 1996); Pascal Boyer, *Religion Explained: The Evolutionary Origins of Religious Thought* (New York: Basic, 2001); and Scott Atran, *In Gods We Trust: The Evolutionary Landscape of Religion* (New York: Oxford University Press, 2002).

9. Stephen Prothero, *God Is Not One: The Eight Rival Religions That Run the World* (New York: HarperOne, 2010), 11–12.

10. Prothero's *God Is Not One*, for example, is typical of most contemporary religious scholarship in showing no interest in looking at religion as a cosmic phenomenon. In its unawareness of the cosmos, postmodern thought remains thoroughly modern.

11. This expression, one that I cite several times later on, is that of physical chemist and avowed materialist Peter W. Atkins, *The 2nd Law: Energy,*

Chaos, and Form (New York: Scientific American Books, 1994), 200. It expresses succinctly the materialist assumption that nothing truly new can ever come into the universe over the course of time.

Chapter Two. Awakening

Epigraph: Ralph Waldo Emerson, "Circles," *Emerson's Essays* (New York: Harper Perennial, 1981), 226.

1. Alfred North Whitehead, *Science and the Modern World* (New York: Free Press, 1925), 191–92. The term "hopeless" in this excerpt does not mean "despairing" but rather "endless" in the sense of never fully exhausted.

2. That the texture of "heaven" is somehow specified and given its identity by what happens in the unfinished story of the universe is an intuition that runs throughout the writings of both Pierre Teilhard de Chardin and Alfred North Whitehead. In this book I want to make the same point, though in a different way, drawing on and adapting ideas developed over many years of teaching and studying the following works of Pierre Teilhard de Chardin: *Writings in Time of War*, trans. René Hague (New York: Harper and Row, 1968); *The Heart of Matter*, trans. René Hague (New York: Harvest, 2002); *The Divine Milieu* (New York: Harper and Row, 1962); *Human Energy*, trans. J. M. Cohen (New York: Harvest/Harcourt Brace Jovanovich, 1962); *The Future of Man*, trans. Norman Denny (New York: Harper and Row, 1964); *How I Believe*, trans. René Hague (New York: Harper and Row, 1969); *Activation of Energy*, trans. René Hague (New York: Harcourt Brace Jovanovich, 1970); *The Human Phenomenon*, trans. Sarah Appleton-Weber (Portland, Ore.: Sussex Academic, 1999). I also adapt to this book's arguments some of the ideas of Alfred North Whitehead as presented in the following works: *Science and the Modern World* (New York: Free Press, 1925); *Adventures of Ideas* (New York: Free Press, 1967); *Modes of Thought* (New York: Free Press, 1968); *Process and Reality*, corrected ed., ed. David Ray Griffin and Donald W. Sherburne (New York: Free Press, 1968).

3. Whitehead, *Science and the Modern World*, 192.

4. For a handy presentation of the new scientific cosmic story, once again, I recommend Brian Swimme and Mary Evelyn Tucker's *Journey of the Universe* (New Haven: Yale University Press, 2011).

5. Daniel Dennett, as interviewed in John Brockman, *The Third Culture* (New York: Touchstone, 1995), 187.

6. In *Resting on the Future: Catholic Theology for an Unfinished Universe* (New York: Bloomsbury, 2015), I used the term "archaeology" as equivalent to what I am calling archaeonomy here.

7. For a recent example see Stephen Cave, "There's No Such Thing as Free Will, but We're Better Off Believing in It Anyway," *Atlantic* (June 2016),

http://www.theatlantic.com/magazine/archive/2016/06/theres-no-such
-thing-as-free-will/480750/.

8. The analogical way might also be called archetypal or participatory, as
    in Platonic and neo-Platonic thought. My concern is to distinguish the
    prescientific participatory understanding of nature from the anticipatory
    vision, represented mainly by significant strains of Abrahamic religion, a
    perspective that I consider especially consonant with new scientific dis-
    coveries. I develop this distinction at book length in *Resting on the Future.*

9. See, for example, Huston Smith, *Why Religion Matters: The Fate of the
    Human Spirit in an Age of Disbelief* (New York: HarperCollins, 2001).

10. For a runaway version of analogical physics see the recent book by Max
    Tegmark, *Our Mathematical Universe: My Quest for the Ultimate Nature of
    Reality* (New York: Knopf, 2014). A more moderate version is that of Alan
    Lightman, *The Accidental Universe: The World You Thought You Knew* (New
    York: Pantheon, 2014). An interesting but still problematic reaction to
    the currently fashionable analogical, Platonic trend in physics appears in
    the recent book by physicist Lee Smolin and Roberto Unger, *The Singu-
    lar Universe and the Reality of Time: A Proposal in Natural Philosophy* (Cam-
    bridge: Cambridge University Press, 2014). These coauthors accept the
    reality of time and have a rudimentary sense of the narrative quality of
    nature. They still read the universe, however, with the same materialist
    eyes as other archaeonomic naturalists.

11. Throughout this book I use the expressions "more-being" and "fuller-
    being" as these appear in the works of Pierre Teilhard de Chardin cited
    earlier. In opposition to both the analogical and archaeonomic readings,
    anticipation claims that what is going on in the universe can be fully un-
    derstood not by looking solely to the temporal past or the eternal present
    but only by looking also up ahead.

12. Terrence Deacon, in his recent book *Incomplete Nature: How Mind Emerged
    from Matter* (New York: Norton, 2012), seems to be aware of the anticipa-
    tory character of nature, and he rightly struggles to find a worldview that
    can accommodate the fact of emergence. In the end, however, he fails to
    break out of the archaeonomic frame of mind that he shares with most
    other scientific thinkers today. For a study of the emergence from a theo-
    logical point of view see Philip Clayton, *Mind and Emergence: From Quan-
    tum to Consciousness* (New York: Oxford University Press, 2006).

13. If some readers have any doubts about my personal affection for contem-
    porary evolutionary biology, they may find it interesting to know that I
    have been honored with a Friend of Darwin award by the National Cen-
    ter for Science Education. It is partly because of the high regard I have
    for the natural sciences that I refuse to accept the facile but trendy confla-
    tion of evolutionary biology with materialist metaphysics. For a defense of
    this point see my book *God after Darwin: A Theology of Evolution,* 2nd ed.
    (Boulder, Colo.: Westview, 2008).

14. Teilhard, *Activation of Energy*, 99–139.

15. Smolin and Unger in *Singular Universe* seem to have partially grasped the problem of archaeonomy's inability to account for emergent phenomena. They try to get around this problem by simply asserting that emergent novelty is "evoked" as the cosmic narrative moves along. However, they provide no metaphysical foundation for the assumption that evocation can be an explanation of the more-being that enters into an emergent universe over the course of time. A more robust explanation of the arrival of more-being is needed. An anticipatory worldview, I am arguing, can account for the more that comes into the world in emergence, and at the same time, as we shall see, it can justify the confidence we need to have in our minds if they are to work at all.

16. This is a paraphrase of the Marxist philosopher Ernst Bloch, who defines the biblical God as having future "as his very essence"; *The Principle of Hope*, trans. Neville Plaice, Stephen Plaice, and Paul Knight (Oxford: Basil Blackwell, 1986), 1:236. This citation is from Jürgen Moltmann, *The Coming of God*, trans. Margaret Kohl (Philadelphia: Fortress, 1975), 23.

17. Emerson, "Circles," 226.

18. See Tomáš Halík, *Patience with God: The Story of the Zacchaeus Continuing in Us* (New York: Random House, 2009).

19. Classic examples of the analogical approach in Western theology are the works of Augustine of Hippo and Thomas Aquinas. A more recent example is David Bentley Hart's learned book *The Experience of God: Being, Consciousness, Bliss* (New Haven: Yale University Press, 2013). For the latter and most other contemporary apologists for analogy, the new scientific cosmic story seems to be of little spiritual or intellectual relevance, not a drama to be read with both religious interest and epistemological patience.

20. This, it seems to me, explains the reluctance of even so insightful a scholar as Seyyed Hossein Nasr to embrace Darwinian biology, which he (in silent agreement with many contemporary evolutionists) considers to be inseparable from archaeonomic materialism. See, for example, his *Knowledge and the Sacred* (Albany: SUNY Press, 1989).

21. An earlier attempt at something like "big history," one that looked inside the temporal world, was Augustine of Hippo's classic work *The City of God* (completed in 426 CE). Augustine, however, approached the task from a predominantly analogical perspective, though intermingled at times with anticipatory elements from the Bible. Of course, he had no sense of deep cosmic time as science does today.

22. Teilhard, *Activation of Energy*, 238. The best study of Teilhard's own sometimes controversial understanding of non-Christian religions is that of Ursula King, *Teilhard de Chardin and Eastern Religions: Spirituality and Mysticism in an Evolutionary World*, rev. ed. (New York: Paulist, 2011).

Chapter Three. Transformation

Epigraph: J. B. S. Haldane, cited at http://en.wikiquote.org/wiki/J._B._S. _Haldane.

1. Simon Blackburn, "Thomas Nagel: A Philosopher Who Confesses to Finding Things Bewildering," *New Statesman* (Nov. 8, 2012), http://www .newstatesman.com/culture/culture/2012/11/thomas-nagel-philosopher -who-confesses-finding-things-bewildering.

2. Charles Darwin, "Letter to W. Graham, July 3rd, 1881," *The Life and Letters of Charles Darwin*, ed. Francis Darwin (New York: Basic, 1959), 285. Similarly, the late philosopher Richard Rorty has remarked: "The idea that one species of organism is, unlike all the others, oriented not just toward its own increased prosperity [that is, toward 'fitness'] but toward Truth, is as un-Darwinian as the idea that every human being has a built-in moral compass—a conscience that swings free of both social history and individual luck"; Richard Rorty, "Untruth and Consequences," *New Republic* (July 31, 1995), 32–36. I first came across the references here to Darwin and Rorty in an online essay entitled "Darwin, Mind, and Meaning" (1996) by Alvin Plantinga. The essay can now be found at https://www.calvin.edu/academic/philosophy/virtual_library/articles/ plantinga_alvin/darwin_mind_and_meaning.pdf.

3. Ian McEwan, *Saturday* (Toronto: Knopf, 2005), 56.

4. A notable exception is the recent book by Thomas Nagel, *Mind and Cosmos: Why the Materialist Neo-Darwinian Conception of Nature Is Almost Certainly False* (New York: Oxford University Press, 2012). Typical reactions by professional philosophers to Nagel's reservations about materialism are a combination of disbelief and ridicule. In my view, however, few recent events in the intellectual world have more openly demonstrated the chokehold that archaeonomic materialism has on contemporary philosophy, and especially on the philosophy of mind, than the intemperate reactions by other scholars to the legitimate questions raised by Thomas Nagel about the intelligibility and truth of evolutionary materialism.

5. This is the gist, for example, of Richard Dawkins's *Climbing Mount Improbable* (New York: Norton, 1996).

6. For closer examination of the inadequacy of materialist accounts see my books *Deeper than Darwin: The Prospect for Religion in the Age of Evolution* (Boulder, Colo.: Westview, 2003); and *Is Nature Enough? Meaning and Truth in the Age of Science* (Cambridge: Cambridge University Press, 2006).

7. Nagel, *Mind and Cosmos*.

8. I have been led to this line of questioning by the writings of the philosopher Bernard Lonergan, especially his impressive book *Insight: A Study of Human Understanding*, 3rd ed. (New York: Philosophical Library, 1970).

9. Bernard Lonergan, *Method in Theology* (New York: Herder and Herder, 1972), 235–37.

10. In the New Testament, St. Paul refers to the work of God in Christ as *dikaiosyne*, a Greek term that means "making things right." This meaning is made more explicit in the German word for justification, *Rechtfertigung*.

11. Lee Smolin, "A Naturalist Account of the Limited, and Hence Reasonable, Effectiveness of Mathematics in Physics" (2015), http://fqxi.org/data/essay-contest-files/Smolin_FQXi_2015_math_final.pdf.

12. For an example of the unwarranted assumption that understanding means simplification, see Alex Rosenberg's *The Atheist's Guide to Reality: Enjoying Life without Illusions* (New York: Norton, 2012), 20–21. Rosenberg is by no means alone in assuming that simplification is identical to understanding. The physicist and author Sean Carroll has repeated the dubious assumption by archaeonomic naturalists that there is anything more fundamental to the universe than what physics can find. See his recent book *The Big Picture: On the Origins of Life, Meaning, and the Universe Itself* (New York: Dutton, 2016).

13. Peter W. Atkins, *The 2nd Law: Energy, Chaos, and Form* (New York: Scientific American Books, 1994), 200; Rosenberg, *Atheist's Guide to Reality*.

14. David Papineau, *Philosophical Naturalism* (Cambridge, Mass.: Blackwell, 1993), 3.

15. Rosenberg, *Atheist's Guide to Reality*, 20–21.

16. As I mentioned earlier, vestiges of the analogical worldview are also present in contemporary mathematical physics insofar as it refuses to take time seriously and instead gives the status of reality only to the timeless realm of mathematical ideas. See especially Max Tegmark, *Our Mathematical Universe: My Quest for the Ultimate Nature of Reality* (New York: Knopf, 2014).

Chapter Four. Interiority

Epigraph: Pierre Teilhard de Chardin, *The Human Phenomenon*, trans. Sarah Appleton-Weber (Portland, Ore.: Sussex Academic, 1999), 6.

1. A clear example of this reductive atomism is Daniel Dennett's celebrated book *Darwin's Dangerous Idea: Evolution and the Meaning of Life* (New York: Simon and Schuster, 1995).

2. David Christian, *Maps of Time: An Introduction to Big History* (Berkeley: University of California Press, 2004). See also Cynthia Stokes Brown, *Big History: From the Big Bang to the Present* (New York: New Press, 2007); Eric Chaisson, *Epic of Evolution: Seven Ages of the Cosmos* (New York: Columbia University Press, 2007); Loyal Rue, *Everybody's Story: Wising Up to the Epic of Evolution* (Albany: State University of New York Press, 2000); Harold Morowitz, *The Emergence of Everything: How the World Became Complex* (Oxford: Oxford University Press, 2002); and Fred Spier, *Big History and the Future of Humanity* (Oxford: Wiley-Blackwell, 2010).

3. Exceptions to an exclusively outside approach can be found in the writ-

ings of Thomas Berry. Berry's book *The Dream of the Earth* follows Pierre Teilhard de Chardin in highlighting subjectivity as an essential aspect of nature. Influenced by both Teilhard and Berry, Brian Swimme and Mary Evelyn Tucker's *Journey of the Universe* (New Haven: Yale University Press, 2011) acknowledges the inside story, as does Duane Elgin's book *The Living Universe* (Oakland, Calif.: Berrett-Koehler, 2009). Such narrative readings of the cosmic story from the inside are rare.

4. Pierre Teilhard de Chardin, *The Human Phenomenon*, trans. Sarah Appleton-Weber (Portland, Ore.: Sussex Academic, 1999).

5. David Christian, "From Mapping to Meaning," in *Creation Stories in Dialogue: The Bible, Science, and Folk Traditions*, ed. Alan Culpepper and Jan G. Van Der Watt (Boston: Brill, 2016), 36.

6. Yuval Noah Hariri, for instance, repeatedly refers to religious ideas as illusions incompatible with scientific understanding: *Sapiens: A Brief History of Humankind* (Toronto: McClelland and Stewart, 2014).

7. E. O. Wilson, *Consilience: The Unity of Knowledge* (New York: Knopf, 1998). Wilson, in this and other books, persistently confuses chronological priority with metaphysical primacy. His *Consilience* professes to unify the cosmic and human story, but it does so only by reducing them both to what can be in principle specified by physical science. Wilson's work is as relentlessly archaeonomic in its method and assumptions as one could possibly imagine.

8. Daniel Dennett, as interviewed in John Brockman's *The Third Culture* (New York: Touchstone, 1995), 187.

9. See, for example, Daniel Dennett, *Breaking the Spell: Religion as a Natural Phenomenon* (New York: Viking, 2006); also, Robert Hinde, *Why Gods Persist: A Scientific Approach to Religions* (New York: Routledge, 1999).

10. This irony is highlighted by Thomas Nagel, *Mind and Cosmos: Why the Materialist Neo-Darwinian Conception of Nature Is Almost Certainly False* (New York: Oxford University Press, 2012). A lifelong materialist himself, Nagel has been honest enough recently to admit that contemporary materialist "explanation" contributes little to an in-depth understanding of mind, especially since it either ignores inwardness altogether or treats it as epiphenomenal. Yet Nagel provides no robust metaphysical alternative to materialism as the default intellectually acceptable worldview. I suspect that for him an anticipatory metaphysics would fall outside the range of contemporary intellectual plausibility. That it may do so, however, would be indicative of the tenacious grip archaeonomy—an internally incoherent worldview—still has on contemporary intellectual life.

11. David Chalmers, "Facing Up to the Problem of Consciousness," *Journal of Consciousness Studies* 2 (1995): 200–219. See also Colin McGinn, *The Mysterious Flame: Conscious Minds in a Material World* (New York: Basic, 1999); for a typical hardline defense of the materialist view of mind, see Owen Flanagan, *The Problem of the Soul: Two Visions of Mind and How to Reconcile Them* (New York: Basic, 2002).

12. For an excellent study of futile attempts by materialist scientists to make sense of subjectivity see David Ray Griffin, *Unsnarling the World-Knot: Consciousness, Freedom, and the Mind-Body Problem* (Berkeley: University of California Press, 1998).

13. For an extended attempt to do so, see my book *Resting on the Future: Catholic Theology for an Unfinished Universe* (New York: Bloomsbury, 2015).

14. For example, John R. Searle, *Mind: A Brief Introduction* (Oxford: Oxford University Press, 2004), 135–36; Steven Pinker, *The Blank Slate: The Modern Denial of Human Nature* (New York: Penguin, 2002); Owen Flanagan, *The Problem of the Soul: Two Visions of Mind and How to Reconcile Them* (New York: Basic, 2002); and Daniel C. Dennett, *Consciousness Explained* (New York: Little Brown, 1991).

15. Alfred North Whitehead, *Modes of Thought* (New York: Free Press, 1968), 156.

16. Whitehead, *Modes of Thought*, 148–69.

17. Alan B. Wallace, *The Taboo of Subjectivity: Toward a New Science of Consciousness* (New York: Oxford University Press, 2000).

18. Hans Jonas, *The Phenomenon of Life* (New York: Harper and Row, 1966), 9–10.

19. Steven Pinker, "The Stupidity of Dignity," *New Republic* (May 28, 2008), https://newrepublic.com/article/64674/the-stupidity-dignity.

20. For a book-length development of this point see my *Is Nature Enough? Meaning and Truth in the Age of Science* (Cambridge: Cambridge University Press, 2006).

21. Most prominent of the names on this list will be that of Pierre Teilhard de Chardin (1881–1955), whom the materialist philosopher Daniel Dennett, for example, refers to as a "loser" (*Darwin's Dangerous Idea*, 320). Dennett's archaeonomic point of view, however, is itself incoherent, at least as a worldview, since it claims that minds, including the minds of its defenders, are ultimately nothing more than the mindless and lifeless stuff to which they believe science has now reduced the universe. As I have already indicated, defenders of materialism cannot help trusting their own minds in the very act of claiming that they are right and others wrong, yet their own archaeonomic worldview cannot justify the trust they have in those minds.

Chapter Five. Indestructibility

1. Alfred North Whitehead, *Science and the Modern World* (New York: Free Press, 1925), 191–92.

2. For this idea and what follows in the present section of this chapter I am indebted to Alfred North Whitehead, *Process and Reality*, corrected ed., ed. David Ray Griffin and Donald W. Sherburne (New York: Free Press, 1968), 29, 34–51, 60, 81–82, 86–104, 340–51; and Alfred North White-

head, "Immortality," in *The Philosophy of Alfred North Whitehead*, ed. Paul A. Schilpp (Evanston: Northwestern University Press, 1941), 682–700; see also Charles Hartshorne, *The Logic of Perfection* (LaSalle, Ill.: Open Court, 1962), 250, 24–62.

3. Hartshorne, *Logic of Perfection*, 24–62.

4. On the human instinct to conquer time by recovering the eternity of beginnings, see Mircea Eliade, *Myth and Reality* (New York: Harper and Row, 1968).

5. The physicist John Archibald Wheeler (1911–2008) is often credited with being the source of the saying that "time is nature's way to keep everything from happening at once," but he admits that he is not its originator. See an account of his remarks online at https://en.wikiquote.org/wiki/John_Archibald_Wheeler.

6. See Thomas Nagel, "What Is It Like to Be a Bat?" *Philosophical Review* 83 (1974): 435–50.

7. On the significance of striving as an essential feature of life see Michael Polanyi, *Personal Knowledge: Towards a Post-Critical Philosophy* (New York: Harper and Row, 1958), 327, 344; and Michael Polanyi, "Life's Irreducible Structure," in *Knowing and Being*, ed. Marjorie Grene (Chicago: University of Chicago Press 1969), 225–39.

Chapter Six. Transcendence

Epigraph: Pierre Teilhard de Chardin, *Writings in Time of War*, trans. René Hague (New York: Harper and Row, 1968), 27–28 [emphasis original].

1. Michael Gazzaniga, *Human: The Science behind What Makes Us Unique* (New York: HarperCollins, 2008), 3.

2. Recall, for example, how the writings of Rosenberg, Blackburn, Dennett, and other materialist philosophers I have cited are content to unify the world by reducing it to its past physical elements and suppressing any search for dramatic unity and narrative coherence ahead.

3. Here I am indebted to ideas of the philosopher Michael Polanyi as set forth, for example, in *The Tacit Dimension* (Garden City, N.Y.: Doubleday Anchor, 1967) and in Michael Polanyi and Harry Prosch, *Meaning* (Chicago: University of Chicago Press, 1975).

4. Eric Chaisson, *Epic of Evolution: Seven Ages of the Cosmos* (New York: Columbia University Press, 2007).

5. On the "striving" of living beings and the "logic of achievement" that differentiates life from nonlife see Michael Polanyi, *Personal Knowledge: Towards a Post-Critical Philosophy* (New York: Harper and Row, 1958), esp. 327–80.

6. See, for example, Owen Flanagan, *The Problem of the Soul: Two Visions of Mind and How to Reconcile Them* (New York: Basic, 2002); Natalie Angier, "My God Problem—and Theirs," *American Scholar* 72 (2004): 131–34;

and Michael Shermer, *How We Believe: The Search for God in an Age of Science* (New York: Freeman, 2000).

7. Heinz Pagels, *Perfect Symmetry* (New York: Bantam, 1985), xv.

8. For additional comments on how science eliminates mystery see the references I have already made to Atkins, Blackburn, Wilson, Rosenberg, Papineau, and Dennett.

9. See Sean Carroll's online article, "Does the Universe Need God?" (n.d.), at http://preposterousuniverse.com/writings/dtung/.

10. For a more extended study of both sober and sunny naturalism see my book *Is Nature Enough? Meaning and Truth in the Age of Science* (Cambridge: Cambridge University Press, 2006). A good example of sober cosmic pessimism is physicist Steven Weinberg's thoughtful book *Dreams of a Final Theory* (New York: Pantheon, 1992).

11. A good example of sunny archaeonomic naturalism is E. O. Wilson's book *Consilience: The Unity of Knowledge* (New York: Knopf, 1998).

12. *Siksasamuccaya*, as adapted from *The Buddhist Tradition*, ed. William Theodore de Bary (New York: Vintage, 1972), 84–85.

13. Excerpted from Michael Harter, *Hearts on Fire: Praying with the Jesuits* (Chicago: Loyola University Press, 2005), 102.

Chapter Seven. Symbolism

Epigraph: Rainer Maria Rilke, *Letters to a Young Poet*, trans. Reginald Snell (Mineola, N.Y.: Dover, 2002), 36.

1. In this respect my approach differs from that of Paul Tillich who, in my opinion, is a predominantly analogical theologian.

2. Paul Tillich, *Theology of Culture*, ed. Robert C. Kimball (New York: Oxford University Press, 1959), 131–32.

3. I owe much of this chapter's understanding of faith and symbolism to the many works of theologian Paul Tillich. The clearest introduction to his understanding of both faith and symbol is his *Dynamics of Faith* (New York: Harper Torchbooks, 1958). Unlike Tillich, however, I am placing my discussion of religious symbolism in the context of our new sense of a still-emerging universe.

4. Dao De Jing, chapter 14, in *Tao: A New Way of Thinking*, trans. and commentary by Chang Chung-yuan (New York: Harper and Row, 1975), 23.

5. Dao De Jing, chapter 11, in *Tao: A New Way of Thinking*, 21.

6. Cited by Chang Chung-yuan in *Tao: A New Way of Thinking*, 36.

7. Ironically, however, an implicit faith, a sense of being grasped by rightness, is operative unnoticed in every scientist's quest for right understanding. It is also present paradoxically in every archaeonomic assertion that materialism is true and religion pure fiction.

8. This belief has become most explicit recently in the writings of the New Atheist manifestoes of Richard Dawkins, *The God Delusion* (New York:

Houghton Mifflin, 2006); Sam Harris, *The End of Faith: Religion, Terror, and the Future of Reason* (New York: Norton, 2004) and *Letter to a Christian Nation* (New York: Knopf, 2007); Daniel Dennett, *Breaking the Spell: Religion as a Natural Phenomenon* (New York: Viking, 2006); and Christopher Hitchens, *God Is Not Great: How Religion Poisons Everything* (New York: Hachette, 2007). See also Victor J. Stenger, *God: The Failed Hypothesis: How Science Shows That God Does Not Exist* (Amherst, N.Y.: Prometheus, 2007); Carl Sagan, *The Demon-Haunted World: Science as a Candle in the Dark* (New York: Ballantine, 1997); Steven Weinberg, *Dreams of a Final Theory* (New York: Pantheon, 1992); Michael Shermer, *How We Believe: The Search for God in an Age of Science* (New York: Freeman, 2000); and Owen Flanagan, *The Problem of the Soul: Two Visions of Mind and How to Reconcile Them* (New York: Basic, 2002).

9. Shermer, *How We Believe.*

10. See Mircea Eliade, *Myth and Reality* (New York: Harper and Row, 1963); and Ian McEwan, *Saturday* (Toronto: Knopf, 2005), 56.

11. For example, Paul M. Churchland, *The Engine of Reason, the Seat of the Soul: A Philosophical Journey into the Brain* (Cambridge: MIT Press, 1995); and Daniel Dennett, *Consciousness Explained* (New York: Little, Brown, 1991).

12. Again, see, for example, the Duke philosopher Alex Rosenberg, *The Atheist's Guide to Reality: Enjoying Life without Illusions* (New York: Norton, 2012), 20–21.

13. See Henri Frankfort et al., *The Intellectual Adventure of Ancient Man: An Essay on Speculative Thought in the Ancient Near East* (Chicago: University of Chicago Press, 1977).

14. See Paul Ricoeur in *The Philosophy of Paul Ricoeur*, ed. Charles Reagan and David Stewart (Boston: Beacon, 1978), 213–22.

15. Rilke, *Letters to a Young Poet*, 36.

16. The New Atheist Christopher Hitchens (in *God Is Not Great*) bases his atheism mostly on Freud's now quaint theory of religion as the infantile projection of a "father figure" onto the heavens, whereas Richard Dawkins (*The God Delusion*) bases his, at least in part, on a Darwinian understanding of life and religion.

Chapter Eight. Purpose

Epigraph: *The Complete Works of Ralph Waldo Emerson* (Boston: Houghton, Mifflin, 1904), 10:155.

1. E. D. Klemke, "Living without Appeal" in *The Meaning of Life*, ed. E. D. Klemke (New York: Oxford University Press, 1981), 169–72; Stephen Jay Gould, *Ever since Darwin: Reflections in Natural History* (New York: Norton, 1977), 12–13.

2. Richard Dawkins, *River out of Eden: A Darwinian View of Life* (New York: Basic, 1995), 96.

3. Hans Jonas, *The Phenomenon of Life* (New York: Harper and Row, 1966), 9–10.

4. E. A. Burtt, *The Metaphysical Foundations of Modern Science* (Garden City, N. Y.: Doubleday Anchor, 1954).

5. Jonas, *The Phenomenon of Life*, 9–10.

6. Jonas, *The Phenomenon of Life*, 9–10. The theologian Paul Tillich also adopts the expression "ontology of death" to characterize the modern sense of the metaphysical primacy of the inorganic over the organic; *Systematic Theology* (Chicago: University of Chicago Press, 1963), 3:19.

7. Christian de Duve, *Vital Dust* (New York: Basic, 1995).

8. This is one of the main arguments of Pierre Teilhard de Chardin's *The Human Phenomenon*, trans. Sarah Appleton-Weber (Portland, Ore.: Sussex Academic, 1999).

9. Jacques Monod, *Chance and Necessity: An Essay on the Natural Philosophy of Modern Biology*, trans. Austryn Wainhouse (New York: Vintage, 1972).

10. As examples readers may look once again at the citations made earlier to writings of Simon Blackburn, David Papineau, Daniel Dennett, Ian McEwan, and Alex Rosenberg, among others.

11. Martin Rees, *Just Six Numbers: The Deep Forces That Shape the Universe* (New York: Basic, 2000); and *Our Cosmic Habitat* (Princeton: Princeton University Press, 2001).

12. What I referred to earlier as "analogical physics" also endorses cosmic pessimism, at least implicitly, by denying the reality of irreversible time, a prerequisite of narrative meaning. See, for example, Max Tegmark, *Our Mathematical Universe: My Quest for the Ultimate Nature of Reality* (New York: Knopf, 2014); and Alan Lightman, *The Accidental Universe: The World You Thought You Knew* (New York: Pantheon, 2014).

13. See for, example, Andrei Linde, "Inflationary Cosmology and the Question of Teleology," in *Science and Religion in Search of Cosmic Purpose*, ed. John F. Haught (Washington, D.C.: Georgetown University Press, 2000) 1–17.

14. Pierre Teilhard de Chardin calls it the "analytical illusion" in *Activation of Energy*, trans. René Hague (New York: Harcourt Brace Jovanovich, 1970), 139.

15. Pierre Teilhard de Chardin, *The Heart of Matter*, trans. René Hague (New York: Harcourt Brace Jovanovich), 49, 122, 131, 226–38.

16. I want to emphasize once again that an anticipatory vision is not at all opposed to scientific method and discovery. It even allows for the predicted "heat death" of our cosmos billions of years from now. However, the anticipatory vision does stand in stark opposition to the materialist archaeonomic vision according to which what is most real and intelligible can be found only by laying out the prebiotic and unconscious elements in earlier chapters of cosmic history.

17. Such a demand is at the forefront, for example, of the writings of the New Atheists listed earlier.

18. Emphasis on beauty as the end of all things is fundamental to Pope Francis's ecological vision as expressed in his recent encyclical *Laudato Si'*: "At the end, we will find ourselves face to face with the infinite beauty of God (cf. 1 Cor 13:12), and be able to read with admiration and happiness the mystery of the universe, which with us will share in unending plenitude." Encyclical Letter *Laudato Si'* no. 243, http://w2.vatican.va/content/francesco /en/encyclicals/documents/papa-francesco_20150524_enciclica-laudato -si.html.

19. This aesthetic vision is set forth most clearly in the works of Alfred North Whitehead. See especially his *Adventures of Ideas* (New York: Free Press, 1967), 252–72, 283–95.

20. A thought often attributed to the Russian novelist Fyodor Dostoyevsky.

21. Whitehead, *Adventures of Ideas*, 62, 183–85, 265.

22. Pierre Teilhard de Chardin, *How I Believe*, trans. René Hague (New York: Harper and Row, 1969), 43–44: "Multiply to your heart's content the extent and duration of progress. Promise the earth a hundred million more years of continued growth. If, at the end of that period, it is evident that the whole of consciousness must revert to zero, *without its secret essence being garnered anywhere at all*, then, I insist, we shall lay down our arms—and mankind will be on strike. The prospect of a *total death* (and that is a word to which we should devote much thought if we are to gauge its destructive effect on our souls) will, I warn you, when it has become part of our consciousness, immediately dry up in us the springs from which our efforts are drawn" [emphasis original].

23. Alfred North Whitehead, *Process and Reality*, corrected ed., ed. David Ray Griffin and Donald W. Sherburne (New York: Free Press, 1968), 345–51.

24. In his relentlessly materialist book *Consilience*, Wilson intellectually and in principle subverts the passionate biophilia (love of life) he tries to cultivate both there and elsewhere. Edward O. Wilson, *Consilience: The Unity of Knowledge* (New York: Knopf, 1998); and Edward O. Wilson, *Biophilia* (Cambridge: Harvard University Press, 1984).

25. This is a point emphasized by Thomas Berry. See his *The Dream of the Earth* (San Francisco: Sierra Club Books, 1988). Berry is a sacramentalist, one who provides a kind of synthesis of the analogical with aspects of the anticipatory approach.

26. I have developed this anticipatory critique of our human destruction of ecological integrity at book length in *The Promise of Nature: Ecology and Cosmic Purpose* (New York: Paulist, 1993). Once again, the notion that ecological abuse goes counter to cosmic destiny is also part of the message of Pope Francis's recent encyclical letter *Laudato Si'* no. 243, http://w2 .vatican.va/content/francesco/en/encyclicals/documents/papa-francesco _20150524_enciclica-laudato-si.html.

Chapter Nine. Obligation

Epigraph: Alfred North Whitehead, *Science and the Modern World* (New York: Free Press, 1967), 191.

1. Immanuel Kant, *Foundations of the Metaphysics of Morals, and What Is Enlightenment?* trans. Lewis W. Beck (New York: Liberal Arts), 1959; Immanuel Kant, *Religion within the Limits of Reason Alone*, trans. Theodore M. Greene and Hoyt H. Hudson (New York: Harper, 1960).

2. In this chapter I am, in part, interpreting what Teilhard de Chardin meant when he wrote: "For the human unit the *initial* basis of obligation is the fact of being born and developing *as a function of a cosmic stream*"; Pierre Teilhard de Chardin, *Human Energy*, trans J. M. Cohen (New York: Harvest/Harcourt Brace Jovanovich, 1962), 29 [emphasis original].

3. For summaries and interpretations of evolutionary accounts of morality, see Robert Wright, *The Moral Animal: Evolutionary Psychology and Everyday Life* (New York: Pantheon, 1994); Matt Ridley, *The Origins of Virtue: Human Instincts and the Evolution of Cooperation* (New York: Penguin, 1998); and Paul Bloom, *Just Babies: The Origins of Good and Evil* (New York: Crown, 2013). See also the sophisticated studies of evolution and altruism in Martin A. Nowak and Sarah Coakley, eds., *Evolution, Games, and God* (Cambridge: Harvard University Press, 2013).

4. For details see George C. Williams, *Adaptation and Natural Selection: A Critique of Some Current Evolutionary Thought* (Princeton: Princeton University Press, 1996); William D. Hamilton, "The Genetical Evolution of Social Behavior," *Journal of Theoretical Biology* 7 (1964): 1–52; John Maynard Smith, *The Evolution of Sex* (New York: Cambridge University Press, 1978); Robert L. Trivers, *Social Evolution* (Menlo Park, Calif.: Benjamin Cummings, 1985); Richard D. Alexander, *Darwinism and Human Affairs* (Seattle: University of Washington Press, 1979). For a more extensive discussion of this chapter's topic see John F. Haught, *Is Nature Enough? Meaning and Truth in the Age of Science* (Cambridge: Cambridge University Press, 2006); and *Deeper than Darwin: The Prospect for Religion in the Age of Evolution* (Boulder, Colo.: Westview, 2003).

5. Pascal Boyer, *Religion Explained: The Evolutionary Origins of Religious Thought* (New York: Basic, 2001), 137–67; Scott Atran, *In Gods We Trust: The Evolutionary Landscape of Religion* (New York: Oxford University Press, 2002).

6. Max Otto, "The Non-Existence of God," in *Issues in Religion*, ed. Allie M. Frazier (New York: Van Nostrand, 1975), 385.

7. Kai Nielsen, "Morality and the Will of God," in *Critiques of God*, ed. Peter Angeles (Buffalo: Prometheus, 1976), 254–56.

8. Friedrich Nietzsche, *The Birth of Tragedy and The Genealogy of Morals*, trans. Francis Golffing (New York: Anchor, 1990).

9. Karl Marx, *Early Writings*, trans. Rodney Livingstone and Gregor Benton

(New York: Penguin Classics, 1992); Sigmund Freud, *The Future of an Illusion*, trans. James Strachey (New York: Norton, 1961).

10. Albert Camus, *The Plague*, trans. Stuart Gilbert (New York: Knopf, 1948), 116–17.

11. Jean-Paul Sartre, *Existentialism Is a Humanism*, trans. Carol Macomber (New Haven: Yale University Press, 1970).

12. Schubert Ogden, *The Reality of God and Other Essays* (New York: Harper and Row, 1977), 120–43.

13. Fyodor Dostoyevsky, *The Brothers Karamazov* (New York: Modern Library, 1950), 721.

14. Again, see Ogden, *The Reality of God*, 120–43.

15. See Pierre Teilhard de Chardin, *Human Energy*, 29.

16. This is a point made in a different context by Schubert Ogden, especially in his book *The Reality of God*, 120–43.

17. I believe this is what the late president of the Czech Republic Václav Havel had in mind when in his late-twentieth-century speeches he sometimes blamed a widespread moral irresponsibility on our having lost the sense that the universe has a purpose.

18. See Teilhard de Chardin, *Human Energy*, 29.

Chapter Ten. Wrongness

1. This is a pervasive motif in the works of Pierre Teilhard de Chardin. See, for example, *Christianity and Evolution*, trans. René Hague (New York: Collins, 1969), 79–95, 131–32.

2. Teilhard, *Christianity and Evolution*, 79–95.

3. Alfred North Whitehead, *Process and Reality*, corrected ed., ed. David Ray Griffin and Donald W. Sherburne (New York: Free Press, 1968), 340–41, 346–51.

4. Pierre Teilhard de Chardin, *Activation of Energy*, trans. René Hague (New York: Harcourt Brace Jovanovich, 1970), 229–44.

5. George Williams, "Mother Nature Is a Wicked Old Witch!" in *Evolutionary Ethics*, ed. Matthew H. Nitecki and Doris V. Nitecki (Albany: State University of New York Press, 1995), 217–31.

6. Stephen J. Gould, Introduction to Carl Zimmer, *Evolution: The Triumph of an Idea from Darwin to DNA* (London: Arrow, 2003), xvi–xvii.

7. Philip Kitcher, *Living with Darwin: Evolution, Design, and the Future of Faith* (New York: Oxford University Press, 2009), 124.

Chapter Eleven. Happiness

Epigraph: Arthur Hugh Clough, "It Fortifies My Soul to Know," http://www .bartleby.com/library/poem/1381.html.

1. The terms translated by the word "right" are *samyañc* (in Sanskrit) and *sammā* (in Pāli).

2. Alfred North Whitehead, *Religion in the Making* (1926; New York: Meridian, 1960), 19.

3. Albert Camus, *The Myth of Sisyphus and Other Essays*, trans. Justin O' Brien (New York: Knopf, 1955).

4. Paul Tillich, *The Courage to Be* (New Haven: Yale University Press, 1952), 78–84.

5. Bernard Lonergan, "Cognitional Structure," in *Collection*, ed. F. E. Crowe (New York: Herder and Herder, 1967), 221–39.

6. See, for example, https://www.youtube.com/watch?v=VofusDy8owA. Dawkins, by the way, has explicitly confirmed his commitment to scientism, for example, in his Tanner Lecture on Human Values at Harvard University, 2003, http://tannerlectures.utah.edu/_documents/a-to-z/d/dawkins_2005.pdf.

7. Clough, "It Fortifies My Soul to Know."

8. In *Breaking the Spell: Religion as a Natural Phenomenon* (New York: Viking, 2006), 21, 55, Daniel Dennett has even used the term "brights" to distinguish and separate scientifically enlightened thinkers from the scientifically uninformed religious masses. His explicit claim that the term "brights" does not imply superiority is hard to believe given the fact that he could have chosen other labels.

9. Bertrand Russell, *Religion and Science* (New York: Oxford University Press, 1961).

10. Bertrand Russell, *A Free Man's Worship and Other Essays* (1917; London: Unwin Paperbacks, 1976), 11–12. For a profound analysis of the influence of the Great War on contemporary cosmic pessimism see Charles A. O'Connor III, *The Great War and the Death of God: Cultural Breakdown, Retreat from Reason, and Rise of Neo-Darwinian Materialism in the Aftermath of World War I* (Washington, D.C.: New Academia, 2014).

11. Jürgen Moltmann, *Theology of Hope: On the Ground and the Implications of a Christian Eschatology*, trans. James W. Leitch (New York: Harper and Row, 1967), 32.

12. Moltmann, *Theology of Hope*, 16.

13. If readers want evidence of the popular persistence of analogy, all they need to do is to attend church services and listen to the typical sermons on Sundays in most suburban Christian churches in the United States.

14. The inclusion of the whole cosmos in the scheme of redemption is already present in the religious thought of the Apostle Paul, especially in Romans 8:19–24.

15. Paul Tillich, *The Shaking of the Foundations* (New York: Scribner, 1948), 151.

Chapter Twelve. Prayerfulness

Epigraph: Pierre Teilhard de Chardin, *The Human Phenomenon*, trans. Sarah Appleton-Weber (Portland, Ore.: Sussex Academic, 1999), 213.

1. All these citations are from Friedrich Heiler, *Prayer*, trans. and ed. Samuel McComb (New York: Oxford University Press, 1932), iv–v. For further development, see my book *What Is Religion?* (New York: Paulist, 1990).

2. Again, for examples, see the earlier quotations from the writings of Rosenberg, Dennett, and Papineau.

3. Paul Tillich, *The Courage to Be* (New Haven: Yale University Press, 1952).

4. Pierre Teilhard de Chardin, *Writings in Time of War*, trans. René Hague (New York: Harper and Row, 1968), 28–29.

5. Teilhard, a most anticipatory scientific and religious thinker, confessed that he too had experienced temperamentally the mystical allure of materialism as a young man and that he had to struggle mightily to surmount it. See Pierre Teilhard de Chardin, *Writings in Time of War*, trans. René Hague (New York: Harper and Row, 1968), 14–71.

6. Einstein's "God" is in many respects like that of the early modern pantheist philosopher Baruch Spinoza (1632–77), as Einstein freely admitted. To Spinoza, God is a necessary being, but since God is said to be indistinguishable from nature, it follows in classical pantheism that the theological attributes of necessity and indestructibility must be transferred from God to nature. Since there is no God who transcends nature, nature itself may be called God. What modern materialism adds to Spinoza's pantheistic monism is the expulsion of interiority or subjectivity from nature as well. See Albert Einstein, *Ideas and Opinions* (New York: Modern Library, 1994).

7. Einstein, *Ideas and Opinions*, 11, 46.

8. Pierre Teilhard de Chardin, *Activation of Energy*, trans. René Hague (New York: Harcourt Brace Jovanovich), 239.

9. Teilhard, *The Human Phenomenon*, 213.

10. Ambrose Bierce, *The Devil's Dictionary* (1903; New York: Dover, 1993), 96.

Index

Abrahamic tradition, 11
altruism, 146
analogical physics, 35, 85, 215n12
analogical religion, 94, 128, 160
analogical spirituality, 132
analogy, 32, 34–41; anticipation and,
    42; *arche* and, 198–99; contempla-
    tion and, 61, 62; cosmic history and,
    61, 164; and cosmic purpose, 154;
    and ecological responsibility, 141;
    epistemological impatience and,
    88; and everlastingness, 81–82;
    evolutionists' attachment to, 170;
    human interiority and, 68; imper-
    ishability and, 86; on love of others,
    155; on morality, 151–53; optimism
    and, 39; physics and, 209n16; prayer
    and, 197–98, 200; religion and, 42,
    61–63, 103, 107; on religion's search
    for happiness, 184, 185, 186; on
    religious symbolism, 117, 121–24;
    rightness and, 63, 111; theological
    worldviews and, 61; time and, 164;
    transformation and, 62; and trust in
    the universe, 157; on wrongness,
    162–65
analysis, and search for understanding,
    136
analytical illusion, 136

analytical vision, 58
anticipation, 32, 35–36, 38–41, 206n11,
    207n15, 215n16; acknowledging
    imperfection of nature, 171;
    analogy and, 42; asceticism of,
    40; cosmos and, 74–75, 77, 128,
    132–33; and denying finality of
    perishing, 83; ecological respon-
    sibility and, 141; evolution and,
    170–71; happiness and, 184–88;
    indestructibility and, 86–88; inside
    cosmic story and, 88–89; inside
    story of, 69; on love of others,
    155–56; on morality, 153–56;
    prayer and, 198–201; on a preliving
    universe, 133; religion and, 20–24,
    42, 43, 64, 82, 91, 103; on religious
    symbolism, 117, 123–25; rightness
    and, 55, 58, 63–64, 111–12; on trans-
    cendence, 107; transformation and,
    63–64; and unity of the universe,
    107–8; wrongness and, 160–63
anticipatory metaphysics, 210n10
anticipatory religion, 94
anxiety: acceptance of, 109; conquering
    of, 4
Aquinas, Thomas, 62
archaeology, 33
archaeonomic illusion, 135–36

archaeonomic materialism, 134: falsity of, 74; philosophy and, 208n4. *See also* archaeonomy

archaeonomic naturalism, 58–59, 67, 75; contradictions within, 141; creed of, 133; and ecological responsibility, 140–41; and the human central nervous system, 129; on immortality, 131–32; impatience of, 112; materialist metaphysics of, 141; on meaning in the cosmos, 128–29; on permanence beyond perishing, 84; subverting cognitive self-confidence, 54; on wrongness, 160. *See also* archaeonomy

archaeonomy, 32–34, 36–41, 215n16; *arche* and, 198–99; assumptions of, 58–59; attraction to, 193–96; Big History and, 41–42, 59; and deities, origins of, 147; on emergence of mind, 50; epistemological impatience and, 88; failures of, 71; grammaticalist fixation of, 100; gratitude and, 192–94; incoherence of, 37; on laws of nature, 130–31; on love of others, 155; as metaphysics of the past, 59–60; missing the inside story, 67–69, 71–72; in modern thought, 186; morality and, 144–48; mundane approach of, 53; and ontology of death, 186; pessimism and, 39, 105–6; prayer and, 191–97, 199–201; on religion, 41–42, 103–7, 147; on religion's search for happiness, 184, 185, 186; on religious symbolism, 117–20, 124; rightness and, 43, 111; on sacramentalism, 122; scriptural literalism and, 70; and sense of cosmic purpose, 154; subjectivity of, 120; symbols of, 118–19; and time, 38, 85; and trust in the universe, 157; on wrongness, 163, 165

arche, 198–99

archetypal vision, 61
astrobiology, 85–86, 133
astrophysics, 33, 133–34
Atlas, 182, 183
atomic physics, 194
atomism, 87
Augustine of Hippo, 62, 198, 207n19, 207n21
Averroes, 62
awakening, 44; drama of, 76; meaning of, 76; religious, 43; symbolism of, 113; waiting and, 108
awareness: emergence of, 50; physical preparations for, 71–72
axial age, 10–12, 44, 96

Barth, Karl, 6
beauty, 216n18; anticipation of, 140; growth of, 156; longing for, 68, 140–41; rightness and, 140–41
Berry, Thomas, 203n2, 209–10n3, 216n25
Bierce, Ambrose, 201
Big Bang, 31, 134–35
Big History, 1–2, 5, 10, 37, 46, 120; archaeonomy and, 41–42, 59, 70; failure of, 69–70; ignoring inside story, 98; religion in, 2–3; science and, 3
Blackburn, Simon, 50–54, 56, 58, 118, 119–20
bodhisattva, 108–9
Bonaventure, 62
Bowker, John, 3, 4, 9
Brahman, 4
Brothers Karamazov, The (Dostoevsky), 152
Buddha, sermons of, 11
Buddhism, 23; enjoyment in, 175; God-consciousness and, 95; religious waiting in, 108–9

Camus, Albert, 150, 177, 179, 182, 183
Carroll, Sean, 105, 209n12

Chaisson, Eric, 99
Chance and Necessity (Monod), 134
Christian, David, 68, 70
Christianity, symbolism in, 112
coherence, search for, 53; narrative,
     40–41, 53, 83, 88–90, 96–102,
     136–38, 161–62, 171–72
consciousness: anticipatory thrust of,
     89; cosmos and, 16. *See also* humans,
     consciousness of
contemplation, spirituality of, 61, 62
conversion, personal, 57
cooperation, imperative toward,
     145–46
Copernicus, 50
cosmic pessimism, 34, 74, 127–32, 154,
     165; happiness and, 179, 181–84;
     ignoring possibility of redemptive
     future, 182; private contentment of,
     182; religion and, 187; sustaining
     of, 177–78
cosmic purpose, 127–41; denial of, 74
cosmic story: insideness of, 3; material-
     ist reading of, 98–99; readings of,
     32–41; religion and, 2, 10, 43;
     rightness and, 43; unfinished, 7
cosmos: analogical reading of, 61;
     anticipation in, 40; anticipatory
     view of, 74–75, 77; awakening of,
     74–75, 123; consciousness and,
     16; created complete, 61; inside
     story of, 15, 32, 68; journey of,
     29; pessimism about, 103–4; and
     religion, 3, 18; self-consciousness
     of, 80
creation stories, 70, 97, 104
Crime and Punishment (Dostoevsky),
     152
culture, mind and, 53–54

Dao de Jing, 11, 115
Daoism, God-consciousness and, 95
dark energy, 69
dark matter, 69

Darwin, Charles, 51, 76, 125, 165–68,
     178, 179
Dawkins, Richard, 118, 119–20,
     178–79, 214n16, 219n6
dawning, 7–25, 44
Deacon, Terrence, 206n12
death: illusion of, 131; ontology of, 72,
     133, 134, 138, 140–41, 186; refusal
     to accept, 79–80; survival of, 131–32
Democritus, 84
Dennett, Daniel, 70, 118, 119–20,
     211n21, 219n8
Descartes, Rene, 132
determinism, 33
Dostoevsky, Fyodor, 152
drama, universe as, 101, 102

ecology, 85–86
ecosystems, destruction of, 140–41
Einstein, Albert, 17, 76, 105, 196, 201
eliminative materialists, 120
emergence, eluding human inquiry, 37
emergent phenomena, 37, 207n15
Emerson, Ralph Waldo, 39
empathy, 169
enjoyment, right vs. wrong, 175
epochs, interlacing of, 1
eternal present, 34, 42, 61
everlastingness: humans connecting
     to, 79–80; religious sense of, 81–84
evil: conquering of, 162; origin of, 162;
     religion and, 177; and transforma-
     tion of the cosmos, 165. *See also*
     suffering
evolution: anticipatory view of, 170;
     components of, 31; rightness of,
     171–72; wrongness and, 165–73
evolutionary fitness, 145
evolutionary naturalists: moral passion
     of, 169; on religion and morality,
     147–49; on religious symbols, 118
exclusivism, 6, 18, 22, 173
extraterrestrial life, 85
extremophiles, 133

faith, anticipatory, 41, 182, 187; and
   being grasped by rightness, 114;
   cosmic future and, 47; expression
   of, 114; patience and, 39, 139;
   protection of, 115; and scientific
   pursuit of truth, 179–80; silence
   and, 115; symbols and, 114
Five Pillars of Islam, 12
Francis (pope), 216n18, 216n26
Freud, Sigmund, 124–25, 150

Gazzaniga, Michael, 97
God: death of, 149, 153; fabrication of,
   105; mystery of, 105; trust in, 109;
   unifying power of, 94
God-consciousness, 95–96, 150;
   archaeonomy's approach to, 106;
   emergence of, 99–100, 190; significance of, 100; skepticism about,
   102–3
God Is Not One (Prothero), 204n10
Gould, Stephen Jay, 129, 167, 169
grammaticalism, 99–102, 122
gratitude, 181, 182, 190, 192–94, 197,
   199

Haldane, J. B. S., 52
happiness, 45; anticipation and, 184–88;
   cosmic pessimism and, 181–84; goal
   of, 176; religion and, 175–88;
   sharing of, 183; truth and, 181
Hartshorne, Charles, 83
heaven, 86
Heiler, Friedrich, 189
Hinduism, 4, 176
Hitchens, Christopher, 214n16
hope, 103, 108
Hopkins, Gerard Manley, 21
humans: anxiety of, 62, 85; central
   nervous system of, 129; connected
   to everlastingness, 79–80; consciousness of, 10, 68, 77, 80;
   contributing to rightness, 28–29;

cooperative tendencies of, 145–46;
   and cosmic purpose, 154; existence
   of, as endless quest, 139; grasped
   by rightness, 93, 114, 181; interior
   lives of, 8–9; inventing morality,
   150; natural prehistory of, 68;
   needing integration, 24; openness
   to rightness, 159–60; religiousness
   of, 3–5, 9; story of, 1–2; subjectivity
   and, 3; vitality of, 177
Hume, David, 53
Huxley, T. H., 165, 167

idolatry, 122, 176
imagery, personal, 113, 114
immortality, belief in, 129, 131–32
impatience, 39, 163, 185; epistemological, 88, 112, 185
imperishability: anticipation and,
   87–88; desire for, 79–80
indestructibility, 44, 79–91
Indian religion, symbolism in, 112
integration, human need for, 24
intelligibility: future of, 42; search for,
   53. *See also* narrative coherence
interiority (inwardness), 44; analogical
   reading of, 68; archaeonomic
   physicalism and, 72–73; emergence
   of, 3, 69, 74. *See also* subjectivity
Israel, prophets of, 11–12

Jaspers, Karl, 10–11, 12, 44. *See also*
   axial age
Jesus, 12, 57, 112, 143, 183; and prayer,
   189–90
Jonas, Hans, 72, 133, 134
Journey of the Universe (Swimme and
   Tucker), 203n1
justification, 160

Kant, Immanuel, 7–8, 143, 144–45
Kitcher, Philip, 167, 169
Klemke, E. D., 129

language, symbols and, 9
Laozi, 11
law, metaphor of, 201
life: drama of, 101–2; emergence of, 38, 43, 66; epoch of, 30; journey of, 156; origin of, 33
light, symbolism of, 113
literalists, 30–31
love, meaning of, 155–56
Luther, Martin, 189

Mahayana Buddhism, 108–9
Maimonides, 62
Maps of Time: An Introduction to Big History (Christian), 68
Marx, Karl, 149–50
materialism, 3, 34, 37, 51, 106; appeal of, 193–96; archaeonomic, 37; on cosmic history, 67; dogmatic, 71; ethical record of, 152; on mind, 76; as ontology of death, 133; on permanence, 84; on religion, 70; rightness and, 57; self-contradictory nature of, 52; as worldview, 119–20
matter: epoch of, 30; hostility of, to life, 134; imperishability of, 84; mindlessness of, 132; re-eternalizing, 84–85; as symbol, 119
McEwan, Ian, 52, 119. See also *Saturday* (McEwan)
meaning, dawning of, 42
mediocrity, principle of, 50–51
metaphor, 93
metaphysics of the future, 88, 199
mind: attracted to rightness, 55; birth of, 76; confidence in, 51–55, 60, 62, 211n21; culture and, 53–54; emergence of, 38, 43, 50–53; epoch of, 30; materialist version of, 51–53; passion of, 97; reduced to mindlessness, 52–54
mindlessness, final state of, 195, 196
modernity, in the cosmic story, 72

Moltmann, Jürgen, 184
Monod, Jacques, 134, 135
monotheism, appeal of, 94
moral aspiration, 3, 18, 68, 149–54
moral duty, 45
morality: archaeonomy and, 144–48; aspiration toward, 149, 154–55; invented by humans, 150; religion and, 143–54; rightness and, 152; in an unfinished universe, 153–56. See also obligation
Muhammad, 12
multiverse, 50–51, 85, 96, 135, 137–38
mystery, 5, 9, 40, 94, 104; acknowledging, 5
myth, 93, 97

Nagel, Thomas, 53, 208n4, 210n10
narrative coherence, 40–41, 53, 83, 88–90, 96–98, 102, 125, 136–38, 161–62, 171–72
narratives, irreversibility of, 83
Nash, Ogden, 84
Nasr, Seyyed Hossein, 62, 207n20
naturalism, 34. See also scientific naturalism
natural science, striving for simplicity, 66–67
natural selection, 165, 166–70
nature: analogical reading of, 62; anticipatory view of, 82, 201; interiority of, 70, 131; laws of, 130; materialist reading of, 37, 67; mindlessness of, 76–77; re-eternalizing, 85; sacramental function of, 164; subjectivity excluded from, 113; unfinished state of, 64; unfolding of, 38; wrongness in, 165–70. See also universe
New Atheists, 5, 12–13, 117–18, 150. See also Dawkins, Richard; Dennett, Daniel; Hitchens, Christopher
Nielsen, Kai, 149

Nietzsche, Friedrich, 149, 153, 183
Noble Eightfold Path, 11, 23
Novalis, 189

objectivism, 74–75
objectivity, myth of, 73
obligation, 45; human sense of, 147; importance of, 145; Marx on, 150. *See also* morality
Origin of Species, On the (Darwin), 168
Otto, Max, 149
oughtness, sense of, 45, 144, 146

Pagels, Heinz, 104–5
pantheism, 195–96, 220n6
panvitalism, 131
Papineau, David, 59–60, 118, 119–20
patience, 39, 40, 42, 139, 162, 176–77
perennialism, 18–19, 22–23, 152
Perennial Philosophy, 19
permanence, 84–85
physicalism, archaeonomic, 72–73
physics: analogical worldview and, 209n16; completeness of, 59–60; Platonic trend in, 206n10
Pinker, Steven, 72
Plague, The (Camus), 150
Plato, 11, 62
postmodern pluralism, 20, 24
prayer: analogy and, 197–98, 200; anticipation and, 198–201; archaeonomy and, 191–97, 199–201; cosmic meaning of, 191–200; as definitional to religion, 189–90; gratitude and, 190, 192–94, 197, 199; petitionary, 193–94, 197–99; in an unfinished universe, 201–2
prayerfulness, 45
preaxial religion, 13
projection, 124–25
Prothero, Stephen, 204n10
psychoanalysis, 124–25, 150
purpose, 45, 128, 139. *See also* cosmic purpose

Rees, Martin, 134–35, 137
relativism, 21, 22, 153
religion: academic interpretations of, 41; analogy and, 42, 61–63, 107; anticipation and, 20–24, 42, 43, 64, 76, 91, 163, 187; archaeonomy and, 41–42, 103–7; attraction of, to rightness, 144, 200; awakening of, 28, 176–77; banishment of, 15; Big History and, 2–3; commonalities in, 6; cosmic approach to, 15, 18, 75, 76; cosmic breakthrough of, 56; in cosmic history, 10, 13–14, 43, 65–66, 81; cosmic significance of, 3, 24–25, 80, 81, 100; criticism of, 57, 75–76; dark side of, 29; dawning of, 56; debunking of, 5, 14; defeatist attitudes and, 149; degeneration of, 177; and deliverance from suffering, 4; emergence of, 8–10, 43–44, 66–67, 69, 75, 190; everlastingness and, 81–84; evil and, 5, 15, 81; evolution and, 19–20; exclusivist approach to, 18, 22, 173; as fiction, 34; function of, 156; happiness and, 175–88; inseparable from the universe, 15; interiority of, 66; intolerability of, 149; language of, 93, 112–14; literalism and, 30–31; materialist view of, 70; meaning of, 42; moral criticism of, 150; morality and, 143–54; mystical turn in, 10, 11; new understanding of, 45; ongoing project of, 109–10; origin of, 70–71; perennialist approach to, 18–19, 22–23; perfectionism and, 173; personal conversion and, 57; pluralism and, 20, 24; prayer and, 189–90; preliterate, 121–22; preparations for, 15; as projection, 124–25; responding to death, 79–81; restrictive thinking about, 14; rightness and, 56–58; role of, in human life, 3–4; scholars estranged from,

16; science and, 10, 45–46; skepticism about, 102–3; symbolism of, 9, 112–13, 117–24; transformation and, 49; unfinished, 5, 7, 15, 163, 173; in an unfinished universe, 5–6; unifying principle of, 13; unity of, 110; as the universe, 43; waiting in, 108–9; wrongness and, 30–31, 159

religious awakening, 43. *See also* awakening

religious consciousness, wrongness and, 163–64

religious conversion, cosmic transformation and, 65

religious experience, 2, 3, 5, 182

religious longing, 177

religious restlessness, 68

religious subjectivity, 2, 17

religious teachers, attributes of, 182–83

religious traditions: tolerance between, 21; unity of, 18–24

religious unity, approaches to, 18–24

revelation, primordial, 19

right enjoyment, 182

rightness: abstract language for, 114; analogy and, 63; anticipation and, 28–29, 55, 58, 63–64, 82, 110, 163; attraction to, 12; awakening of, 82; awakening the cosmos, 123; awakening to, 27, 36, 40, 57, 95–96; beauty and, 140–41; common enjoyment of, 183; conforming to, 176; control of, 114–15; cosmic story and, 43; dawning of, 11–13, 15, 42–43, 54, 55, 115, 168–72, 190; detachment and, 62; effectiveness of, 116; encounter with, 57–58; evidence of, 42–43; expectation of, 20–21; grasping of, 55–57, 114, 181; hiddenness of, 111–12; horizon of, 55–57; human consciousness and, 159–60; indestructible, 77, 80–83, 86; infinite extension of, 93; influence of, 115–16; intuition of,

159; materialism and, 57; mind attracted to, 55; morality and, 152; names for, 57, 114; permanence of, 44; personal imagery for, 113, 114; priority of, over religion, 149; proper names for, 113; pursuit of, 178; reality of, 56–57, 151; realization of, 128, 139, 154; redemptive, 23; relationship to, 121; religion and, 56–58, 100, 144, 154, 176–77, 200; revelation of, 122; scientific objectification of, 139; self-revelation of, 44; skepticism about, 102–3; striving for, 102; symbols of, 112–13; transcendence and, 44–45; understanding of, 94; unlimited, 21; victory of, 160; worship of, 27–28

right understanding, 178. *See also* truth

Rilke, Rainer Maria, 124

rituals, 9, 93, 113, 121

Roman Catholicism, analogical spiritualities of, 39

Rorty, Richard, 208n2

Rosenberg, Alex, 60, 118, 119–20, 209n12

Rothe, Richard, 189

Russell, Bertrand, 182, 183–84

sacramentalism, 107, 197

sacraments, 121–22

Sagan, E. O., 85

salvation, 102, 107

Sartre, Jean-Paul, 150, 153

Saturday (McEwan), 52, 53, 54, 56, 58, 119

Schleiermacher, Friedrich, 189

scholars, estranged from religion, 16

science: advance of, 50–51; constraints of, 68–69; detachment of, 2; expectations for, 37; faith in, 50; freedom and, 196; maturation of, 76; opposition to, 21; outside perspective of, 32; pessimism and, 194; religion and, 45–46; trust in, 60

scientific materialism, 3, 23–24; confidence of, 56; glorification of, 51
scientific method, 16, 73, 195
scientific naturalism, 19–20, 23–24, 34,
75–76
scientific truth, faith assumptions of,
179–80
scientism, 179
secularism, contemporary, 122–23
sexuality, 122
Shermer, Michael, 118
signs, distinct from symbols, 112
silence, 115, 122
Sisyphus, 177, 179, 182
Smith, Huston, 62
Smith, William Cantwell, 203n7
Smolin, Lee, 58, 206n10, 207n15
Socrates, 11
soul, 82, 83
Spinoza, Baruch, 220n6
storytelling, 29, 30, 97, 98–99, 168
striving, 90, 101–2, 110
subjectivity, 2, 3, 9, 15, 77; denial of,
71–74; emergence of, 16, 46, 50,
90; evolution and, 169; hidden
world of, 69; illusion of, 120;
insistence on, 75; materialists'
approach to, 89–90; in nature, 131;
obliteration of, 84; and opening of
the universe, 132; origin of, 16;
power of, 16; reduced to material
processes, 71; study of, 66; transcendental goals and, 91; undeni-
ability of, 16; understanding, 71–72;
universe linked with, 16. *See also*
interiority (inwardness)
suffering, 4, 161. *See also* evil;
wrongness
symbolism, 45, 76, 112–13
symbols, 9, 93; faith and, 114; meaning
of, 115; projection and, 124–25;
religion and, 9, 106, 112; understanding of, 116

Teilhard de Chardin, Pierre, 46–47,
109, 203n2, 210, 211n21, 216n22,
220n5
theology, classical, 152
thought, emergence of, 14, 66
Tillich, Paul, 213n1, 213n3
time, slow passage of, 67
timelessness, 35
tissue, saving of, 85
transcendence, 44–45; analogical vision
of, 103; anticipatory vision of, 103,
107; archaeonomy's approach to,
103–6; drawing toward, 93–94;
inexhaustibility of, 94; meaning of,
93–94; openness to, 110; personal
symbols of, 114; power of, 94;
presupposition of, 110; reality of,
151; silence and, 115; skepticism
about, 102–5; trust in, 103–4;
unifying, 99–100. *See also* unity
transformation, 44, 49; analogy and,
62; anticipation and, 63–64; cosmic,
60; inside story of, 69; and religion's
arrival, 76; religious, 65, 121
trust, 162
truth: attraction to, 55; happiness and,
181; pursuit of, 178–80; as value,
178–79

Unger, Roberto, 206n10, 207n15
unhappiness, threat of, 176
unity: analogical reading of, 95;
anticipatory reading of, 95, 98;
archaeonomic reading of, 94, 98;
discovery of, 95–98; fact of, 94–95;
primacy of, over plurality, 95;
principle of, 13; striving for, 102;
transcendent, 95, 100
universe: analogical reading of, 34–41;
anticipatory reading of, 35–36,
38–41, 54, 63, 100; archaeonomic
reading of, 32–34, 36–41; awakening of, 8, 42, 67, 100; beauty in,

140–41; beginning of, 8; complexity in, 59; destiny of, 188; dramatic quality of, 101, 102; as endless quest, 139; energy flow in, 99; fixed, 7–8, 17–18; hidden meaning in, 177, 178; human trust in, 156–57; impermanence of, 84–85; inside story of, 1–2, 83; intelligibility of, 62, 77, 137; materialist picture of, 50–51; meaning of, 15, 40–41; meaninglessness of, 57, 74, 99, 128, 194; mindlessness of, 16–17, 132; narrative character of, 11, 83, 136–38; ongoing development of, 14–15; perishability of, 139; pointlessness of, 14, 60; purpose of, 132; self-consciousness of, 97–100; self-unifying transitions of, 96–100; story of, 29–32; subjects in, 2; transformation of, 49; unfinished, 5–6, 7, 153–56. *See also* nature

Upanishads, 4, 11

virtue, cultivation of, 156
vitality, 110, 132, 177

waiting, 107–8. *See also* patience
Wheeler, John Archibald, 212n5
Whitehead, Alfred North, 28, 29, 40, 81, 82, 176
Williams, George, 167, 169
Wilson, E. O., 70, 85, 118, 119–20, 141, 210n7, 216n24
worship, 2, 5, 144, 149, 190
wrongness, 45; analogical view of, 163–65; anticipatory view of, 160–63; archaeonomic view of, 163, 165; attraction to, 175; awareness of, 159; evolution and, 165–73; existence of, 162; moral, 161; natural, 161; religion and, 159; religious consciousness and, 163–64; temporality and, 164; victory of, 160, 163

Wu Cheng, 116